ENERGY POLICIES, POLITICS AND PRICES

U.S. ENERGY INFRASTRUCTURE

CLIMATE CHANGE VULNERABILITIES AND ADAPTATION EFFORTS

ENERGY POLICIES, POLITICS AND PRICES

Additional books in this series can be found on Nova's website under the Series tab.

Additional e-books in this series can be found on Nova's website under the e-book tab.

ENERGY POLICIES, POLITICS AND PRICES

U.S. ENERGY INFRASTRUCTURE

CLIMATE CHANGE VULNERABILITIES AND ADAPTATION EFFORTS

JOANNE R. BALLARD
EDITOR

New York

Copyright © 2015 by Nova Science Publishers, Inc.

All rights reserved. No part of this book may be reproduced, stored in a retrieval system or transmitted in any form or by any means: electronic, electrostatic, magnetic, tape, mechanical photocopying, recording or otherwise without the written permission of the Publisher.

For permission to use material from this book please contact us:
nova.main@novapublishers.com

NOTICE TO THE READER

The Publisher has taken reasonable care in the preparation of this book, but makes no expressed or implied warranty of any kind and assumes no responsibility for any errors or omissions. No liability is assumed for incidental or consequential damages in connection with or arising out of information contained in this book. The Publisher shall not be liable for any special, consequential, or exemplary damages resulting, in whole or in part, from the readers' use of, or reliance upon, this material. Any parts of this book based on government reports are so indicated and copyright is claimed for those parts to the extent applicable to compilations of such works.

Independent verification should be sought for any data, advice or recommendations contained in this book. In addition, no responsibility is assumed by the publisher for any injury and/or damage to persons or property arising from any methods, products, instructions, ideas or otherwise contained in this publication.

This publication is designed to provide accurate and authoritative information with regard to the subject matter covered herein. It is sold with the clear understanding that the Publisher is not engaged in rendering legal or any other professional services. If legal or any other expert assistance is required, the services of a competent person should be sought. FROM A DECLARATION OF PARTICIPANTS JOINTLY ADOPTED BY A COMMITTEE OF THE AMERICAN BAR ASSOCIATION AND A COMMITTEE OF PUBLISHERS.

Additional color graphics may be available in the e-book version of this book.

Library of Congress Cataloging-in-Publication Data

ISBN: 978-1-63482-286-2

Published by Nova Science Publishers, Inc. † New York

CONTENTS

PREFACE

This book examines what is known about potential impacts of climate change on U.S. energy infrastructure; measures that can reduce climate-related risks and adapt energy infrastructure to climate change; and the role of the federal government in adapting energy infrastructure and adaptation steps selected federal entities have taken.

Chapter 1 - According to the NRC and the USGCRP, changes in the earth's climate—including higher temperatures, changes in precipitation, rising sea levels, and increases in the severity and frequency of severe weather events—are under way and expected to grow more severe over time. These impacts present significant risks to the nation's energy infrastructure.

Economic losses arising from weather-related events—including floods, droughts, and storms—have been large and are increasing, according to USGCRP. Adaptation—an adjustment to natural or human systems in response to actual or expected climate change—is a risk-management strategy to help protect vulnerable sectors and communities that might be affected by climate change.

GAO was asked to examine the vulnerability of the nation's energy infrastructure to climate change impacts. This report examines: (1) what is known about potential impacts of climate change on U.S. energy infrastructure; (2) measures that can reduce climate-related risks and adapt energy infrastructure to climate change; and (3) the role of the federal government in adapting energy infrastructure and adaptation steps selected federal entities have taken. GAO reviewed climate change assessments; analyzed relevant studies and agency documents; and interviewed federal agency officials and industry stakeholders, including energy companies at four sites that have implemented adaptive measures.

Chapter 2 – Changes in the global climate system are unmistakable, as is now evident from observations of increased global average air and ocean temperatures, decreased historical snow pack, rising global average sea level, and more frequent severe weather events. The U.S. Department of Energy (DOE) recognizes that changes in the global climate system could have a profound impact on the Department's mission activities. DOE is committed to reducing greenhouse gas (GHG) emissions and mitigating climate change by developing clean energy and energy efficiency technologies for commercial deployment while providing leadership through its own sustainable operations. As effects of climate change are felt across the world, it is necessary to characterize potential impacts on the DOE mission, programs, and operations to foster adaptation and resilience. DOE will identify where to focus resources to develop greater resilience over time, minimize potential risks and maximize potential opportunities created by climate change. It is important to note that while longer term damages could be very substantial, we may have more modest, nearer term impacts (above and beyond those that we already have due to weather vulnerabilities). The 2014 DOE Climate Change Adaptation Plan (Adaptation Plan) outlines Departmental vulnerabilities and serves to guide our response to allow DOE to continue to achieve its mission.

The DOE vision for climate change adaptation is the integration of risk based resiliency to address identified climate change vulnerabilities across all DOE programs and policies wherever appropriate. Assessment of climate change vulnerabilities, informed by best available science, are seen as an integral part of the DOE's planning activities, risk assessment, and careful investment that define the DOE mission.

Climate change adaptation is not new to DOE; rather, climate change is an ongoing part of DOE research, modeling, and policy development. A strong culture of preparedness, integrated safety management, and operational excellence in potentially hazardous working environments already exists throughout DOE. Climate change resilience will build on this operational and capital planning, as well as provide information to the larger applied research body of climate change adaptation.

Chapter 3 - Though climate change science often is portrayed as controversial, broad scientific agreement exists on many points:

- The Earth's climate is warming and changing.
- Human-related emissions of greenhouse gases (GHG) and other pollutants have contributed to warming observed since the 1970s and,

if continued, would tend to drive further warming, sea level rise, and other climate shifts.

- Volcanoes, the Earth's relationship to the Sun, solar cycles, and land cover change may be more influential on climate shifts than rising GHG concentrations on other time and geographic scales. Human-induced changes are super-imposed on and interact with natural climate variability.
- The largest uncertainties in climate projections surround feedbacks in the Earth system that augment or dampen the initial influence, or affect the pattern of changes. Feedback mechanisms are apparent in clouds, vegetation, oceans, and potential emissions from soils.
- There is a wide range of projections of future, human-induced climate change, all pointing toward warming and associated sea level rise, with wider uncertainties regarding the nature of precipitation, storms, and other important aspects of climate.
- Human societies and ecosystems are sensitive to climate. Some past climate changes benefited civilizations; others contributed to the demise of some societies. Small future changes may bring benefits for some and adverse effects to others. Large climate changes would be increasingly adverse for a widening scope of populations and ecosystems.

As is common and constructive in science, scientists debate finer points. For example, a large majority but not all scientists find compelling evidence that rising GHG have contributed the most influence on global warming since the 1970s, with solar radiation a smaller influence on that time scale. Most climate modelers project important impacts of unabated GHG emissions, with low likelihoods of catastrophic impacts over this century. Human influences on climate change would continue for centuries after atmospheric concentrations of GHG are stabilized, as the accumulated gases continue to exert effects and as the Earth's systems seek to equilibrate.

The U.S. government and others have invested billions of dollars in research to improve understanding of the Earth's climate system, resulting in major improvements in understanding while major uncertainties remain. However, it is fundamental to the scientific method that science does not provide absolute proofs; all scientific theories are to some degree provisional and may be rejected or modified based on new evidence. Private and public decisions to act or not to act, to reduce the human contribution to climate change or to prepare for future changes, will be made in the context of accumulating evidence (or lack of evidence), accumulating GHG concentrations, ongoing debate about risks, and other considerations (e.g., economics and distributional effects).

In: U.S. Energy Infrastructure
Editor: Joanne R. Ballard
ISBN: 978-1-63482-286-2
© 2015 Nova Science Publishers, Inc.

Chapter 1

CLIMATE CHANGE: ENERGY INFRASTRUCTURE RISKS AND ADAPTATION EFFORTS*

United States Government Accountability Office

WHY GAO DID THIS STUDY

According to the NRC and the USGCRP, changes in the earth's climate—including higher temperatures, changes in precipitation, rising sea levels, and increases in the severity and frequency of severe weather events—are under way and expected to grow more severe over time. These impacts present significant risks to the nation's energy infrastructure.

Economic losses arising from weather-related events—including floods, droughts, and storms—have been large and are increasing, according to USGCRP. Adaptation—an adjustment to natural or human systems in response to actual or expected climate change—is a risk-management strategy to help protect vulnerable sectors and communities that might be affected by climate change.

GAO was asked to examine the vulnerability of the nation's energy infrastructure to climate change impacts. This report examines: (1) what is known about potential impacts of climate change on U.S. energy

* This is an edited, reformatted and augmented version of a United States Government Accountability Office publication, No. GAO-14-74, dated January 2014.

infrastructure; (2) measures that can reduce climate-related risks and adapt energy infrastructure to climate change; and (3) the role of the federal government in adapting energy infrastructure and adaptation steps selected federal entities have taken. GAO reviewed climate change assessments; analyzed relevant studies and agency documents; and interviewed federal agency officials and industry stakeholders, including energy companies at four sites that have implemented adaptive measures.

What GAO Found

According to assessments by the National Research Council (NRC) and the U.S. Global Change Research Program (USGCRP), U.S. energy infrastructure is increasingly vulnerable to a range of climate change impacts—particularly infrastructure in areas prone to severe weather and water shortages. Climate changes are projected to affect infrastructure throughout all major stages of the energy supply chain, thereby increasing the risk of disruptions. For example:

- *Resource extraction and processing infrastructure,* including oil and natural gas platforms, refineries, and processing plants, is often located near the coast, making it vulnerable to severe weather and sea level rise.
- *Fuel transportation and storage infrastructure*, including pipelines, barges, railways and storage tanks, is susceptible to damage from severe weather, melting permafrost, and increased precipitation.
- *Electricity generation infrastructure,* such as power plants, is vulnerable to severe weather or water shortages, which can interrupt operations.
- *Electricity transmission and distribution infrastructure,* including power lines and substations, is susceptible to severe weather and may be stressed by rising demand for electricity as temperatures rise.

In addition, impacts to infrastructure may also be amplified by a number of broad, systemic factors, including water scarcity, energy system interdependencies, increased electricity demand, and the compounding effects of multiple climate impacts.

A number of measures exist to help reduce climate-related risks and adapt the nation's energy systems to weather and climate-related impacts. These options generally fall into two broad categories—hardening and resiliency. Hardening measures involve physical changes that improve the durability and stability of specific pieces of infrastructure—for example, elevating and sealing water-sensitive equipment—making it less susceptible to damage. In contrast, resiliency measures allow energy systems to continue operating after damage and allow them to recover more quickly; for example, installing back-up generators to restore electricity more quickly after severe weather events.

In general, the federal government has a limited role in directly adapting energy infrastructure to the potential impacts of climate change, but key federal entities can play important supporting roles that can influence private companies' infrastructure decisions and these federal entities are initiating steps to begin adaptation efforts within their respective missions. Energy infrastructure adaptation is primarily accomplished through planning and investment decisions made by private companies that own the infrastructure. The federal government can influence companies' decisions through providing information, regulatory oversight, technology research and development, and market incentives and disincentives. Key federal entities, such as the Department of Energy, the Environmental Protection Agency, the Federal Energy Regulatory Commission, and the Nuclear Regulatory Commission have also begun to take steps to address climate change risks—through project-specific activities such as research and development and evaluating siting and licensing decisions under their jurisdiction, as well as through broader agency-wide assessments and interagency cooperation.

ABBREVIATIONS

AWF	America's Wetland Foundation
BOEM	Bureau of Ocean Energy Management
BSEE	Bureau of Safety and Environmental Enforcement
CBO	Congressional Budget Office
CEA	Council of Economic Advisers
CEQ	Council on Environmental Quality
CSP	concentrated solar power
DOE	Department of Energy
EIA	Energy Information Administration
ENO	Entergy New Orleans
EPA	Environmental Protection Agency
FEMA	Federal Emergency Management Agency

FERC	Federal Energy Regulatory Commission
FPL	Florida Power and Light
IPCC	Intergovernmental Panel on Climate Change
NCDC	National Climate Data Center
NERC	North American Electric Reliability Corporation
NETL	National Energy Technology Laboratory
NFIP	National Flood Insurance Program
NOAA	National Oceanic and Atmospheric Administration
NRC	National Research Council
NWP	National Water Program
NWS	National Weather Service
NYH	New York Harbor
ONRR	Office of Natural Resources Revenue
PG&E	Pacific Gas and Electric Company
PMA	Power Marketing Administration
TVA	Tennessee Valley Authority
UN	United Nations
USGCRP	U.S. Global Change Research Program

* * *

January 31, 2014

Congressional Requesters

Climate change is a complex, crosscutting issue that could pose significant risks to the nation's energy infrastructure. According to assessments by the National Research Council (NRC)[1] and the United States Global Change Research Program (USGCRP),[2] the effects of climate change are already under way and are projected to continue.[3] Global atmospheric emissions of greenhouse gases have increased markedly over the last 200 years which has contributed to a warming of the earth's climate as well as increasing the acidity of oceans. Changes observed in the United States include more intense weather and storm events, heat waves, floods, and droughts; rising sea levels; and changing patterns of rainfall. These trends, which are expected to continue, can adversely affect energy infrastructure such as natural gas and oil production platforms, pipelines, power plants, and electricity distribution lines, according to NRC and USGCRP, thus making it more difficult to ensure a reliable energy supply to the nation's homes and businesses.

Energy infrastructure can be affected by both acute weather events and long-term changes in the climate, according to NRC and the Department of Energy (DOE). In particular, energy infrastructure located along the coast is at risk from increasingly intense storms, which can substantially disrupt oil and gas production and cause temporary fuel or electricity shortages. In 2012, for example, storm surge and high winds from Hurricane Sandy—an acute weather event-–downed power lines, flooded electrical substations, and damaged or temporarily shut down several power plants and ports, according to DOE, leaving over 8 million customers without power.4 [5] Long-term changes in the climate could also impact energy infrastructure, according to USGCRP and DOE. For example, warming air temperatures may reduce the efficiency of power plants while increasing the overall demand for electricity, potentially creating supply challenges. In addition, while many climate change impacts are projected to be regional in nature, the interconnectedness of the nation's energy system means that regional vulnerabilities may have wide-ranging implications for energy production and use, ultimately affecting transportation, industrial, agricultural, and other critical sectors of the economy that require reliable energy.

As observed by USGCRP, the impacts and financial costs of weather disasters—resulting from floods, drought, and other weather events—are expected to increase in significance as what are historically considered to be "rare" events become more common and intense due to climate change. [6] According to National Oceanic and Atmospheric Administration's (NOAA) National Climate Data Center (NCDC), the United States experienced 11 extreme weather and climate events in 2012, each causing more than $1 billion in losses.[7] Two of the most significant weather events during 2012 were Hurricane Sandy, estimated at $65 billion, and an extended drought that covered over half of the contiguous United States estimated at $30 billion. While it is difficult to attribute any individual weather event to climate change, these events provide insight into the potential climate-related vulnerabilities the United States faces. In this regard, both private sector firms and federal agencies have documented an increase in weather-related losses. A 2013 study by the reinsurance provider Munich Re, for example, indicated that, in 2012, insured losses in the United States totaled $58 billion—far above the 2000 to 2011 average loss of $27 billion. The energy sector often bears a significant portion of these costs, according to USGCRP; for example, direct costs to the energy industry following Hurricanes Katrina and Rita in 2005 were estimated at around $15 billion.[8]

We have reported in the past that policymakers increasingly view climate change adaptation—defined as adjustments to natural or human systems in response to actual or expected climate change—as a risk management strategy to protect vulnerable sectors and communities that might be affected by changes in the climate.[9] State and local governments and the private sector play key roles in planning and implementing energy infrastructure, and some are already engaged in various types of adaptation measures, including vulnerability assessments, strengthening or relocating vulnerable infrastructure, deploying more climate-resilient technologies, and improving electricity grid operations and responsiveness. While some of these measures may be costly, there is a growing recognition that the cost of inaction could be greater. As stated in a 2010 NRC report, increasing the nation's ability to respond to a changing climate can be viewed as an insurance policy against climate risks.[10] To that end, emerging federal efforts are under way to facilitate more informed decisions about adaptation. However, we have reported these federal efforts have been largely carried out in an ad hoc manner, with little coordination across federal agencies or with state and local governments.[11] In 2013, our most recent update to the list of programs at high risk of waste, fraud, abuse, and mismanagement, we identified the federal government's management of climate change risks as an area in need of fundamental transformation due to the fiscal exposure it presents.[12]

In this context, you asked us to examine the vulnerability of the nation's energy infrastructure to climate change. This report examines: (1) what is known about the potential impacts of climate change on U.S. energy infrastructure, (2) measures that can reduce climate-related risks and adapt the energy infrastructure to climate change, and (3) the role of the federal government in adapting energy infrastructure to the potential impacts of climate change, including what steps selected federal entities have taken towards adaptation.

To examine what is known about the impacts of climate change on U.S. energy infrastructure, we reviewed climate change impact assessments from the NRC, USGCRP and federal agencies.[13] We examined potential impacts to the following infrastructure categories, representing four main stages of the energy supply chain:[14] (1) resource extraction and processing infrastructure, (2) fuel transportation and storage infrastructure,(3) electricity generation infrastructure, and (4) electricity transmission and distribution infrastructure. We also assessed broad, systemic factors that may amplify climate change impacts to energy infrastructure.

To identify and examine measures that can reduce climate-related risks and adapt energy infrastructure to climate change, we analyzed relevant studies and government reports and interviewed knowledgeable stakeholders including representatives from professional associations such as the American Gas Association and the National Association of State Energy Officials. We identified and selected a nonprobability sample of four energy companies where decision makers were taking steps to adapt their energy infrastructure to the potential impacts of climate change: Colonial Pipeline Company, Entergy Corporation, Florida Power and Light Company, and Pacific Gas and Electric Company.

To select our sample we conducted a literature review and interviewed officials from research organizations, such as the Environmental and Energy Study Institute and the Center for Climate and Energy Solutions. Our sample selection reflects a range of geographic locations and climate-related risks, as well as infrastructure used in three of the four stages of the energy supply chain.[15]

To examine the role of the federal government in adaptation and steps selected federal entities have taken, we identified federal agencies with key responsibilities related to energy infrastructure by reviewing relevant literature, including previous GAO reports, and interviewing agency officials and knowledgeable stakeholder groups.[16] We compiled an initial list of 15 federal entities that had a connection to energy infrastructure and then narrowed the list to the five that have the most direct influence on energy infrastructure adaptation decisions: DOE, the Environmental Protection Agency (EPA), the Federal Energy Regulatory Commission (FERC), the Nuclear Regulatory Commission as well as the North American Electric Reliability Corporation (NERC).[17] Appendix I provides a more detailed description of our objectives, scope, and methodology.

We conducted this performance audit from July 2012 to January 2014 in accordance with generally accepted government auditing standards. Those standards require that we plan and perform the audit to obtain sufficient, appropriate evidence to provide a reasonable basis for our findings and conclusions based on our audit objectives. We believe that the evidence obtained provides a reasonable basis for our findings and conclusions based on our audit objectives.

BACKGROUND

This section describes: (1) potential climate change impacts in the United States, (2) energy infrastructure in the United States, and (3) climate change adaptation as a risk management tool.

Observed and Projected Climate Change Impacts in the United States

According to assessments by USGCRP, NRC, and others, changes in the earth's climate attributable to increased concentrations of greenhouse gases may have significant environmental and economic impacts in the United States. These changes, summarized in Table 1, involve a wide range of current and projected impacts. While uncertainty exists about the exact nature, magnitude, and timing of climate change, over the next several decades, these and other impacts are projected to continue and likely accelerate, with effects varying considerably by region, according to NRC assessments. Because emitted greenhouse gases remain in the atmosphere for extended periods of time, some changes to the climate are expected to occur as a result of emissions to date, regardless of future efforts to control emissions.[18,19]

Table 1. Current and Projected Climate Changes in the United States

Category	Observed climate changes	Projected climate changes
Temperature	• U.S. average annual temperature has risen about 1.5 degrees Fahrenheit since record keeping began in 1895; more than 80 percent of this increase has occurred since 1980. The most recent decade was the nation's warmest on record. [a] • The frost-free season has been lengthening since the 1980s, and rising temperatures are reducing ice volume on land, lakes, and sea. Minimum Arctic sea ice has decreased by more than 40 percent since satellite records began in 1978.	• U.S. temperatures are expected to continue to rise, with varying impacts by region. In the next few decades, warming of 2 to 4 degrees Fahrenheit is expected for most parts of the nation. • The frost-free season is expected to increase by a month or more and is projected to occur across most of the United States by the end of the century. • Antarctic sea ice is projected to decline in future decades[b,c]

Category	Observed climate changes	Projected climate changes
Precipitation	• Data indicate an overall upward trend in annual precipitation across most of the United States, with an average 5 percent increase since 1900. • Heavy downpours are increasing in most regions of the United States, especially over the last three to five decades.	• Projections of future precipitation indicate that northern areas are expected to continue to become wetter, and southern areas, particularly in the Southwest, are expected to become drier. • Further increases in the frequency and intensity of extreme precipitation events are projected for most U.S. areas.
Sea level rise and coastal erosion	• Global sea level has risen by about 8 inches since reliable record keeping began in 1880. • The current rate of global sea level rise is faster than at any time in the past 2000 years.	• Sea levels are projected to continue to rise, but the extent is not well-understood.[d] • In the next several decades, sea level rise and land subsidence could combine with storm surges and high tides to increase flooding in coastal regions.
Extreme weather events and storms	• Certain types of extreme weather events, such as heat waves, floods, and droughts, have become more frequent and intense in some regions. • In the eastern Pacific, the strongest hurricanes have become stronger since the 1980s, while the total number of storms has declined.	• The intensity of the strongest hurricanes is projected to continue as the oceans continue to warm, causingwind, precipitation, and storm surges. • Other trends in severe storms, including the number of hurricanes and intensity and frequency of tornadoes are uncertain and remain under study.

Sources: GAO analysis of USGCRP's 2009 and 2013 draft National Climate Assessments and NRC's America's Climate Choices: Adapting to the Impacts of Climate Change, 2010.

[a] A report by the United Kingdom notes global mean surface temperatures rose rapidly from the 1970s, but have been relatively flat over the most recent 15 years to 2013. This has prompted speculation that human induced global warming is no longer happening, or at least will be much smaller than predicted. Others maintain that this is a temporary pause in global temperatures and that they will again rise at rates seen previously. United Kingdom Met Office, *Observing Changes in the Climate System: The Recent Pause in Global Warming (1) What do observations of the climate system tell us?* (United Kingdom: July 2013).

[b]U.S. Climate Change Science Program (now known as USGCRP), *Global Climate Change Impacts in the United States*, Draft 2013 National Climate Assessment (Washington, D.C., 2013).

[c]Liu, J, Judith A. Curry, *Accelerated Warming in the Southern Ocean and its Impacts on the Hydrological Cycle and Sea Ice*. School of Earth and Atmospheric Sciences, Georgia Institute of Technology (Atlanta, GA: 2010).

[d] Sea level has been rising, and at an increasing rate, but understanding all of the dynamics involved is not sufficiently complete to allow for an accurate prediction of the likely total extent of sea level rise this century. For example, scientists have a well-developed understanding of the contributions of thermal expansion of the oceans due to warming. However, other changes, such as ice sheet dynamics, are less well-understood , and while this variable is expected to make a significant contribution to sea level rise, quantifying that contribution is difficult.

U.S. Energy Infrastructure

U.S. energy infrastructure comprises four key components: (1) resource extraction and processing infrastructure, such as equipment to extract and refine coal, natural gas and oil; (2) fuel transportation and storage infrastructure, including physical networks of natural gas and oil pipelines; (3) electricity generation infrastructure, including coal-fired, gas-fired, and nuclear power plants, as well as renewable energy infrastructure; and (4) electricity transmission and distribution infrastructure, such as power lines that transport energy to consumers (see fig. 1). According to DOE, the energy supply chain has grown increasingly complex and interdependent. In total, the U.S. energy supply chain includes approximately 2.6 million miles of interstate and intrastate pipelines, 6,600 operational power plants, about 144 operable refineries, and about 160,000 miles of transmission lines. Collectively, this infrastructure enables the United States to meet industrial, commercial, and residential demands, as well as to support transportation and communication networks. The nation's energy supply chain is designed to respond to weather variability, such as changes in temperature that affect load or rapid changes in renewable resource availability that affect supply. These short-term fluctuations are managed by designing redundancy into energy systems and using tools to predict, evaluate, and optimize response strategies in the near term. For example, electrical utilities are beginning to deploy automated feeder switches that open or close in response to a fault condition identified locally or to a control signal sent from another location. When a fault occurs, automated feeder switching immediately reroutes power among distribution circuits isolating only the portion of a circuit where the fault has occurred. This results in a significant reduction in the number of customers affected by an outage and the avoidance of costs typically borne by customers when outages occur, according to a 2013 White House report.[20]

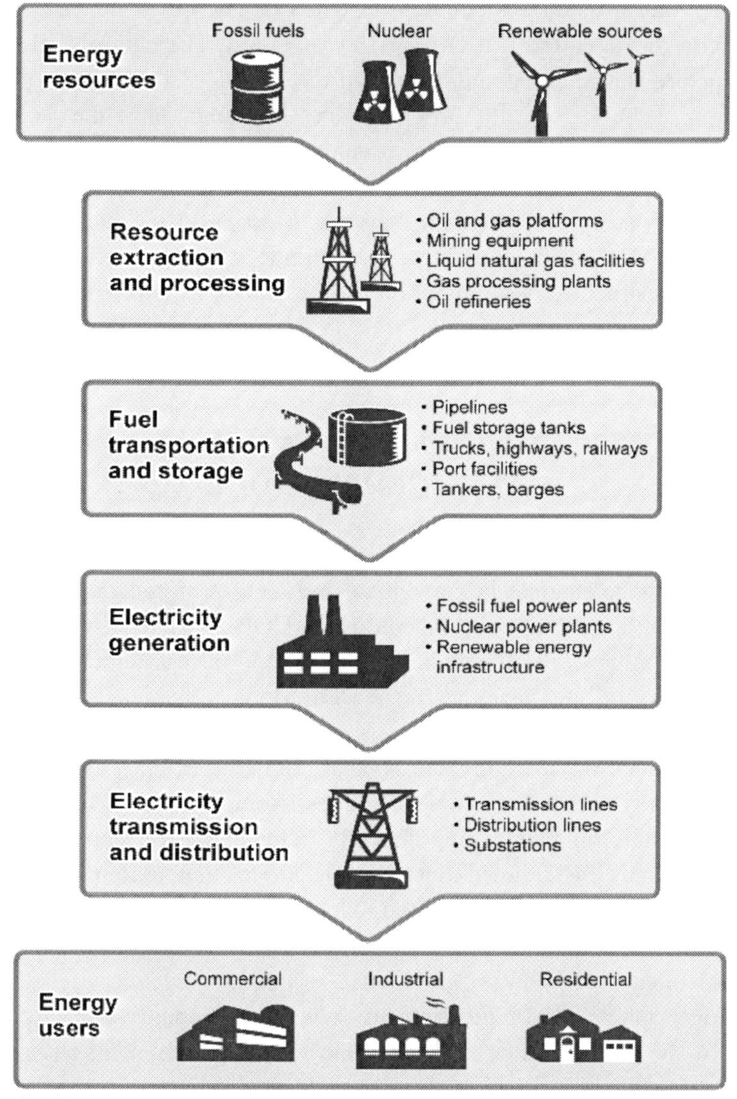

Source: GAO.

Figure 1. Illustration of U.S. Energy Supply Chain.

However, most energy infrastructure was engineered and built for our past or current climate and may not be resilient to continued and expected increases in the magnitude and frequency of extreme weather events and overall continued weather and climate change in the longterm. Further, this infrastructure is aging, according to DOE. For example, most of the U.S. electricity transmission system was designed to last 40 to 50 years; yet, in some parts of the country, it is now 100 years old. The nation's oil and gas infrastructure is also aging and about half of the nation's oil and gas pipelines were built in the 1950s and 1960s. Changes in climate have the potential to further strain these already aging components by forcing them to operate outside of the ranges for which they were designed. DOE reported that aging infrastructure is more susceptible than newer assets to the hurricane-related hazards of storm surge, flooding, and extreme winds, and retrofitting this existing infrastructure with more climate-resilient technologies remains a challenge.

Climate Change Adaptation As a Risk Management Tool

Climate change adaptation addresses the vulnerability of natural and human systems to changes in the climate and focuses on reducing the damage resulting from those changes.[21] According to DOE, two broad ways to reduce the potential impacts of climate change on energy infrastructure are to invest in hardening and resiliency efforts. DOE defines hardening as physical changes to infrastructure to make it less susceptible to storm damage, such as high winds, flooding, or flying debris. DOE defines resiliency as the ability to recover quickly from damage to facilities' components or to any of the external systems on which they depend.[22] The Intergovernmental Panel on Climate Change (IPCC) noted more flexible and resilient systems have greater adaptive capacity and are better suited to handle a changing climate.[23]

Additionally, adaptation requires making policy and management decisions that cut across traditional economic sectors, jurisdictional boundaries, and levels of government. While most energy infrastructure is owned by the private sector, both state and federal governments have roles in energy infrastructure siting, permitting, and regulation. For example, state public utility commissions are responsible for setting the rates for electric service within each state, and owners of energy infrastructure must work with state commissions in order to request rate increases to cover the cost of hardening their infrastructure. Owners of energy infrastructure that spans more than one state, such as natural gas or oil pipelines or electric power lines, may have to work with multiple state commissions on rate and licensing matters and with FERC regarding the rates, terms, and conditions of sales of electricity and transmission in interstate commerce.

U.S. ENERGY INFRASTRUCTURE IS INCREASINGLY VULNERABLE TO A RANGE OF PROJECTED CLIMATE-RELATED IMPACTS

According to USGCRP, NRC, and others, climate change poses risks to energy infrastructure at all four key stages in the supply chain. In addition, broad, systemic factors such as water scarcity and energy system interdependencies could amplify these impacts.

Climate Change Poses Risks to Energy Infrastructure Across the Four Key Stages in the Supply Chain

Impacts from climate change can affect infrastructure throughout the four major stages of the energy supply chain: (1) resource extraction and processing infrastructure, (2) fuel transportation and storage infrastructure, (3) electricity generation infrastructure, and (4) electricity transmission and distribution infrastructure.

Climate Change Can Impact Resource Extraction and Processing Infrastructure

Much of the infrastructure used to extract, refine, and process, and prospect for fuels—including natural gas and oil platforms, oil refineries, and natural gas processing plants—is located offshore or near the coast, making it particularly vulnerable to sea level rise, extreme weather, and other impacts, according to USGCRP and DOE assessments. The Gulf Coast, for example, is home to nearly 4,000 oil and gas platforms (see fig. 2), many of which are at risk of damage or disruption due to high winds and storm surges at increasingly high sea levels.[24] Low-lying coastal areas are also home to many oil refineries, coal import/export facilities, and natural gas processing facilities that are similarly vulnerable to inundation, shoreline erosion, and storm surges. Given that the Gulf Coast is home to approximately half of the nation's crude oil and natural gas production—as well as nearly half of its refining capacity—regional severe weather events can have significant implications for energy supplies nationwide. In 2005, for example, high winds and flooding from Hurricanes Katrina and Rita caused extensive damage to the region's natural gas and oil infrastructure, destroying more than 100 platforms, damaging 558 pipelines, and shutting down numerous refineries, effectively halting nearly all oil and gas production for several weeks. (Fig. 3 illustrates damage to the Mars and Typhoon deepwater platforms following the 2005 hurricanes.) More recently, Hurricane Sandy caused flooding and outages at refineries and petroleum terminals in the New York Harbor area, according to a 2013 DOE report comparing the impacts of northeast hurricanes on energy infrastructure, depressing regional oil supply and leading to temporary price increases.[25]

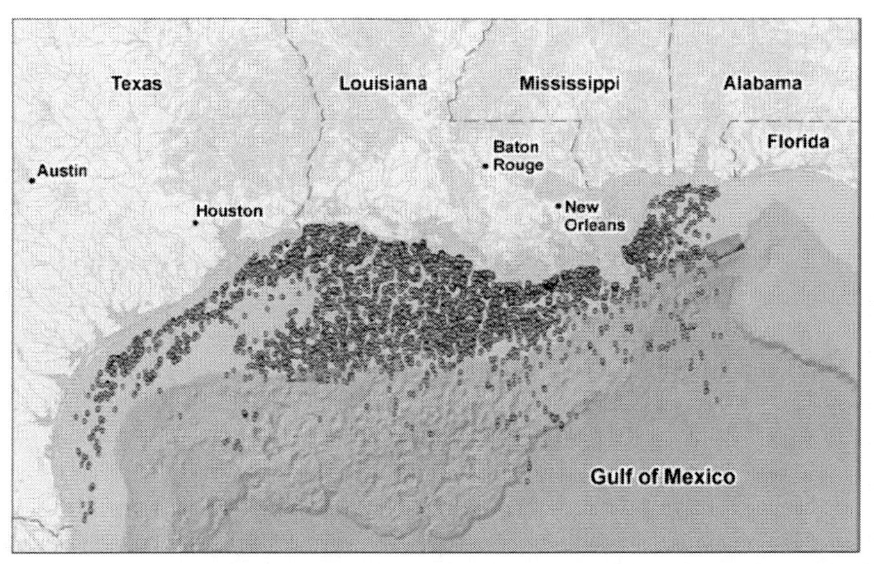

Sources: Based on an online geographic information systems-based mapping tool of the Flower Garden Banks Sanctuary, using data from the National Oceanic and Atmospheric Administration (NOAA) National Marine Sanctuaries Program and NOAA's National Coastal Data Development Center.

Note: Nearly 4,000 active oil and gas platforms are located in the central and western Gulf of Mexico.

Figure 2. Active Oil and Gas Platforms in the Central and Western Gulf of Mexico.

Sources: V. Bhatt, J. Eckmann, W. C. Horak, and T. J. Wilbanks: *Possible Indirect Effects on Energy Production and Distribution in the United States in Effects of Climate Change on Energy Production and Use in the United States*. A report by the U.S. Climate Change Science Program and the Subcommittee on Global Change Research (Washington D. C. 2007). (Mars platform photos); Det Norske Veritas, Technical Report prepared for the Minerals Management Service, *Pipeline Damage Assessment from Hurricanes Katrina and Rita in the Gulf of Mexico,* Report No. 448 14183, January 22, 2007 (Typhoon platform photos).

Figure 3. Damage to Mars and Typhoon Platforms from Hurricanes Katrina and Rita, 2005.

Storm-related impacts on natural gas and oil production infrastructure can also have significant economic implications. Losses related to infrastructure damage can be extensive, particularly given the high value and long life span of natural gas and oil platforms, refineries, and processing plants. For example, a report by Entergy Corporation, an integrated energy company serving a number of southern states, estimated its infrastructure restoration costs at around $1.5 billion following Hurricanes Katrina and Rita. A 2009 DOE assessment reported that some damages resulting from the 2005 hurricanes were too costly to repair; as a result, a number of platforms were sunk, and significant crude oil production capacity was lost.[26] In addition to causing physical damage, increasingly intense severe weather events can disrupt operations and decrease fuel supplies, resulting in broader economic losses for businesses and industries that depend on these resources. According to USGCRP assessments, damage to key infrastructure—especially to refineries, natural gas processing plants, and petroleum terminals—can cause fuel prices to spike across the country, as evidenced by Hurricanes Katrina and Sandy. Flood damage is the most common and costliest type of storm damage to oil production infrastructure, resulting in the longest disruptions, according to DOE's 2010 report.

Warming temperatures and water availability may also present challenges for the nation's extraction and processing infrastructure. For example, according to USGCRP, climate change impacts have already been observed in Alaska, where thawing permafrost has substantially shortened the season during which oil and gas exploration and extraction equipment can be operated on the tundra.[27] Oil refineries around the nation are also potentially at risk, according to USGCRP; they require both significant quantities of water and access to electricity, making them vulnerable to drought and power outages.

Climate Change Can Impact Fuel Transportation and Storage Infrastructure

USGCRP assessments identified several ways in which climate change can affect fuel transportation infrastructure, including pipeline systems that carry natural gas and oil; trucks, railways, and barges that transport coal, oil and petroleum products; as well as storage facilities, such as aboveground tanks, underground salt caverns, and aquifers.[28]

Natural gas and oil pipelines, which generally require electricity to operate, are particularly vulnerable to extreme weather events, according to DOE. The U.S. pipeline system is a complex network comprising over 2.6 million miles of natural gas and oil pipelines, some of which have already been affected by past weather events. For example, electric power outages from Hurricane Katrina caused three critical pipelines— which cumulatively transport 125 million gallons of fuel each day—to shut down for two full days and operate at reduced power for about two weeks, leading to fuel shortages and temporary price spikes. In addition to the power outage, the Department of the Interior's Minerals Management Service reported that approximately 457 pipelines were damaged during the hurricanes, interrupting production for months (see fig. 4).[29] More recently, in July 2011, ExxonMobil's Silvertip pipeline in Montana, buried beneath the Yellowstone riverbed, was torn apart by flood-caused debris, spilling oil into the river and disrupting crude oil transport in the region, with damages estimated at $135 million, according to the Department of Transportation.[30] Storm surge flooding can also affect aboveground fuel storage tanks, according to DOE; for example, tanks not fully filled can drift off of their platforms or become corroded by trapped salt water.

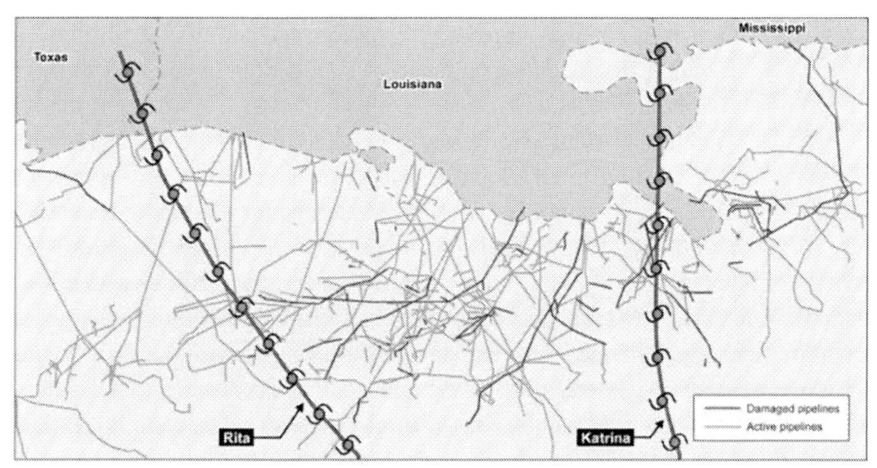

Source: Det Norske Veritas, Technical Report prepared for the Department of the Interior's Minerals Management Service, *Pipeline Damage Assessment from Hurricanes Katrina and Rita in the Gulf of Mexico,* Report No. 448 14183, January 22, 2007.

Figure 4. Pipeline Damages Reported for Hurricanes Katrina and Rita, 2005.

In addition to pipelines, rail, barge, and tanker trucks also play critical roles in transporting fuel across the country. According to USGCRP and DOE assessments, fuel transport by rail and barge can be affected when water levels in rivers and ports drop too low, such as during a drought, or too high, such as during a storm surge. During the 2012 drought, the U.S. Army Corps of Engineers reported groundings of traffic along the Mississippi River due to low water depths, preventing barge shipments of coal and petroleum products. Lower water levels can also affect the amount of fuel the barges are capable of hauling; according to DOE's 2013 assessment, a one-inch drop in river level can reduce a barge's towing capacity by 255 tons.

Fuel transportation infrastructure can also be affected by rising temperatures, according to assessments by DOE and USGCRP. For example, in 2012, Hurricane Sandy's storm surge produced nearly four feet of floodwaters, damaging or temporarily shutting down the Port of New York and New Jersey, as well as electrical systems, highways, and rail track. Disruptions in barge transportation due to extreme weather can also present challenges for areas such as Florida, which are nearly entirely dependent on barges for fuel delivery. Intense storms and flooding can also wash out rail lines—which in many regions follow riverbeds—and impede the delivery of coal to power plants. According to DOE, flooding of rail lines has already been a problem both in the Appalachian region and along the Mississippi River. The rerouting that occurs as a result of such flooding can cost millions of dollars and can delay coal deliveries.

Colder climates present a different set of risks for fuel transportation infrastructure, according to DOE and USGCRP assessments. For example, in Alaska—where average temperatures have risen about twice as much as the rest of the nation—thawing permafrost is already causing pipeline, rail, and pavement displacements, requiring reconstruction of key facilities and raising maintenance costs.[31] Melting sea ice caused by warmer temperatures can result in more icebergs and ice movement, which in turn can damage barges transporting natural gas and oil. On the other hand, decreasing sea ice could also generate some benefits for the natural gas and oil sectors; USGCRP reports that warmer temperatures are expected to improve shipping accessibility in some areas of the Arctic Basin, including oil and gas transport by sea.[32]

Climate Change Can Impact Electricity Generation Infrastructure

According to assessments by USGCRP, DOE, and others, climate change will have a significant impact on the nation's electricity generation facilities, including fossil fuel and nuclear power plants—which together produce the vast majority of the nation's electricity—as well as renewable energy infrastructure such as wind turbines and hydropower dams.

Fossil fuel and nuclear power plants. According to USGCRP, climate change is expected to have potentially significant consequences for fossil fuel and nuclear power plants. Fossil fuel plants—which burn coal, natural gas, or oil—are susceptible to much of the same impacts as nuclear power plants, according to USGCRP and DOE, including diminishing water supplies, warming temperatures, and severe weather, among others.

According to USGCRP, episodic and long-lasting water shortages and elevated water temperatures may constrain electricity generation in many regions of the United States. As currently designed, most fossil fuel and nuclear plants require significant amounts of water to generate, cool, and condense steam. Energy production, together with thermoelectric power, accounted for approximately 11 percent of U.S. water consumption in 2005, according to one study[33], second only to irrigation.[34] Issues related to water already pose a range of challenges for existing power plants, as illustrated by the following examples cited by DOE[35]:

- *Insufficient amounts of water.* In 2007, a drought affecting the southeastern United States caused water levels in some rivers, lakes and reservoirs to drop below the level of intake valves that supply cooling water to power plants, causing some plants to stop or reduce power production.
- *Outgoing water too warm.* In 2007, 2010, and 2011, the Tennessee Valley Authority had to reduce power output from its Browns Ferry Nuclear Plant in Alabama because the temperature of the river was too high to receive discharge water without raising ecological risks; the cost of replacing lost power was estimated at $50 million.[36]
- *Incoming water too warm.* In August 2012, Dominion Resources' Millstone Nuclear Power Station in Connecticut shut down one reactor because the intake cooling water, withdrawn from the Long Island Sound, exceeded temperature specifications. The resulting loss of power production was estimated at several million dollars.

USGCRP and NRC assessments project that water issues will continue to constrain electricity production at existing facilities as temperatures increase and precipitation patterns change. Many of these risks are regional in nature; research by the Electric Power Research Institute (EPRI), for example, indicates that approximately 25 percent of existing electric generation in the United States is located in counties projected to be at high or moderate water supply sustainability risk in 2030.[37] Water availability concerns are already affecting the development of new power plants, according to USGCRP's 2009 assessment, as plans to develop new plants are delayed or halted at increasing rates. Moreover, as demands for energy and water increase, competition between the energy, industrial, and agricultural sectors, among others, sectors could place additional strain on the nation's power plants, potentially affecting the reliability of future electric power generation.[38]

USGCRP and DOE assessments also indicate that higher air and water temperatures may diminish the efficiency by which power plants convert fuel to electricity. A power plant's operating efficiency is affected by the performance of the cooling system, among other things. According to USGCRP, warming temperatures may decrease the efficiency of power plant cooling technologies, thereby reducing overall electricity generation. While the magnitude of these effects will vary based on a number of plant-and site-specific factors, USGCRP assessments suggest that even small changes in efficiency could have significant implications for electricity supply at a national scale. For example, an average reduction of 1 percent in electricity generated by fossil fuel plants nationwide would mean a loss of 25 billion kilowatt-hours per year, about the amount of electricity consumed by approximately 2 to 3 million Americans. When projected increases in air and water temperatures associated with climate change are combined with changes to water availability, generation capacity during the summer months may be significantly reduced, according to DOE. Warmer water discharged from power plants into lakes or rivers can also harm fish and plants; such discharges generally require a permit and are monitored.

In addition to the effects of rising temperatures and reduced water availability, power plant operations are also susceptible to extreme weather, increased precipitation, and sea level rise, according to assessments by USGCRP and DOE. To a large extent, this vulnerability stems from their location—thermoelectric power plants are frequently located along the U.S. coastline, and many inland plants sit upon low-lying areas or flood plains. For coastal plants, more intense hurricane-force winds can produce damaging storm surges and flooding—an impact illustrated by Hurricane Sandy, which shut down several power plants. Some power plants near the coast could also be affected by sea level rise, according to DOE, because they are located on land that is relatively flat and, in some places, subsiding. Increasing intensity and frequency of flooding also poses a risk to inland power plants, according to DOE. The structures that draw cooling water from rivers are vulnerable to flooding and, in some cases, storm surge. This risk was illustrated when Fort Calhoun nuclear power plant was initially shut down for a scheduled refueling outage in April 2011. According to Nuclear Regulatory Commission officials, the outage was subsequently extended due to flooding from the Missouri River and a need to address long-standing technical issues that continued to impair plant operations.[39] According to USGCRP, seasonal flooding could result in increased costs to manage on-site drainage and runoff.

Renewable energy infrastructure. Overall, use of renewable energy is growing in the United States, according to the Energy Information Administration (EIA), with hydropower and wind representing the largest renewable sources of electricity in 2012. Renewable energy sources generally produce much lower emissions of greenhouse gases, the primary anthropogenic driver of climate change. However, these sources can also be affected by climate change, given their dependence on water resources, wind patterns, and solar radiation. Specific impacts to these sectors are described below:

- *Hydropower.* Hydropower—a major source of electricity in some regions of the United States, particularly the Northwest—is highly sensitive to a number of climactic changes. According to USGCRP and DOE, rising temperatures can reduce the amount of water available for hydropower—due to increased evaporation—and degrade habitats for fish and other wildlife. Hydropower production is also highly sensitive to changes in precipitation and river discharge, according to USGCRP and DOE assessments. According to USGCRP's 2009 assessment, for example, studies suggest that every 1 percent decrease in precipitation results in a 2 to 3 percent drop in streamflow; in the Colorado Basin, such a drop decreases hydropower generation by 3 percent. Climate variability has already had a significant influence on the operation of hydropower systems, according to USGCRP, with significant changes detected in the timing and amount of streamflows in many western rivers.
- *Biofuels.* According to USGCRP assessments, biofuels made from grains, sugar and oil crops, starch, grasses, trees, and biological waste are meeting an increasing portion of U.S. energy demand.[40] Currently, however, most U.S. biofuels are produced from corn grown on rain-fed land, making biofuel susceptible to drought and reduced precipitation, as well as competing demands for water.[41] These issues were highlighted when droughts in 2012 produced a poor corn harvest, raising concerns about the allocation of corn for food versus ethanol. Production of biofuel crops may also be inhibited by heavy rainfall and flooding, according to DOE. Climate change could also present some benefits; for example, warmer temperatures could

extend the period of the growing season (although DOE also notes that extreme heat could damage crops).

- *Solar.* The effects of climate change on solar energy—which generated about 0.05 percent of U.S. electricity in 2010—depend on the type of solar technology in use, according to DOE and USGCRP. Some studies suggest that photovoltaic energy production could be affected by changes in haze, humidity, and dust. Higher temperatures can also reduce the effectiveness of photovoltaic electricity generation. On the other hand, concentrating solar power (CSP) systems— unlike photovoltaic cells—require extensive amounts of water for cooling purposes, making them susceptible to water shortages.[42]
- *Wind.* Wind energy accounted for about 13 percent of U.S. renewable energy consumption in 2011, but its use is growing rapidly, according to EIA. Unlike thermoelectric generation, wind energy does not use or consume water to generate electricity, making it a potentially attractive option in light of water scarcity concerns. On the other hand, wind energy cannot be naturally stored, and the natural variability of wind speeds can have a significant positive or negative impact on the amount of energy produced. Wind turbines are also subject to extreme weather, according to USGCRP.
- *Geothermal.* Geothermal power plants extract geothermal fluids—hot water, brines, and steam—from the earth by drilling wells to depths of up to 10,000 feet. According to EIA, geothermal energy represented approximately 2 percent of U.S. energy consumption in 2011, with most geothermal reservoirs located in western states, Alaska, and Hawaii. As with fossil fuel power plants and concentrating solar power, increases in air and water temperatures can reduce the efficiency with which geothermal facilities generate electricity, according to DOE's 2013 assessment. Geothermal power plants can also withdraw and consume significant quantities of water, according to DOE, making them susceptible to water shortages caused by changes in precipitation or warming temperatures.

Climate Change Can Impact Electricity Transmission and Distribution Infrastructure

Transmission and distribution infrastructure can extend for thousands of miles, making it vulnerable to a variety of climate change impacts.[43] According to assessments by USGCRP and others, transmission and distribution lines and substations are susceptible to damage from extreme winds, ice, lightning strikes, wildfires, landslides, and flooding (see fig. 5). High winds, especially when combined with precipitation from tropical storms and hurricanes, can be particularly damaging, potentially interrupting service in broad geographic areas over long periods of time.[44] In the winter months, heavy snowfall[45] and excessive icing on overhead lines can cause outages and require costly repairs, according to a review of literature published in the journal *Energy*.[46] According to USGCRP, increasing temperatures and drought may increase the risk of wildfires, which in turn may cause physical damage to electricity transmission infrastructure and decrease available transmission capacity. Apart from transmission and distribution lines, severe weather can also present risks for substations, according to DOE, which modify voltage for residential and commercial use, as well as for operation centers that are critical components of any electricity supply system.

Source: KOMO News (upper left); National Oceanic and Atmospheric Administration Photo Library, National Weather Service Collection (upper right); Federal Emergency Management Agency/Win Henderson (bottom photos).

Figure 5. Examples of Weather-Related Electrical Grid Disturbances.

Since 2000, there has been a steady increase in the number of weather-related grid disruptions in the United States, according to DOE. These disruptions—which are often a result of trees damaging distribution lines—can result in high costs for utilities and consumers, including repair costs for damaged equipment and economic costs related to work interruptions, lost productivity, and other factors. Some recent events, as reported by DOE, illustrate these vulnerabilities. In 2012, for example, about 3 to 4 million customers lost power due to a combination of thunderstorms and strong winds—known as a derecho—that affected the Midwest and Mid-Atlantic coast. Hurricane Sandy's impact was even more severe, according to DOE, with electricity outages affecting around 8.7 million customers.[47] DOE further reported that around 1.4 million of these customers were still without power 6 days later. Winter conditions can also pose risks to the electrical grid; in February 2013, a winter storm caused extensive damage to transmission systems in the Northeast, causing over 660,000 customers in eight states to lose power. See figure 6 for weather-related grid disruptions.

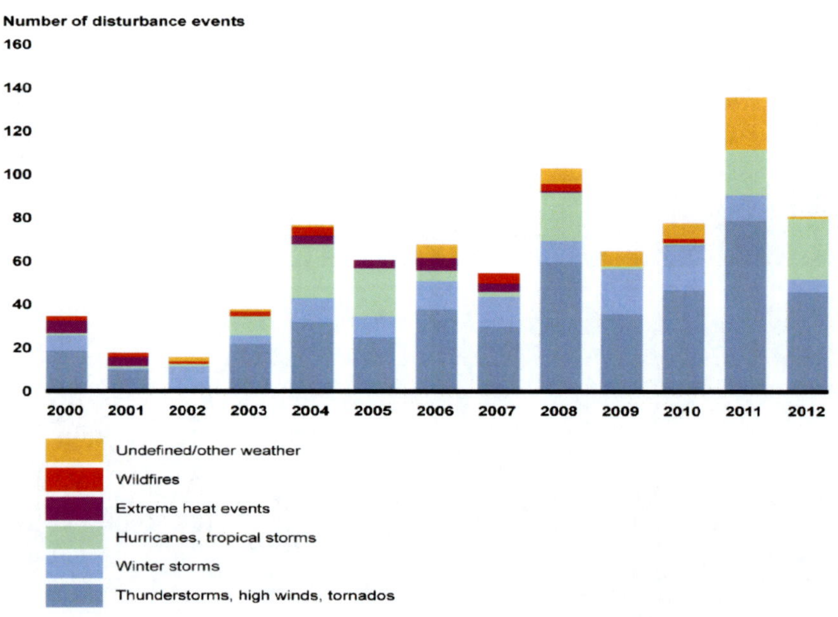

Source: DOE, "Electric: Disturbance Events (OE-417)." U.S. Department of Energy, Office of Electricity Delivery and Energy Reliability, April 2013. http://www.oe.netl.doe.gov/oe417.aspx. Data analysis by Evan Mills, Lawrence Berkeley National Laboratory.

Figure 6. Weather-Related Grid Disruptions, 2000-2012.

Apart from risks related to extreme weather events, increasing temperatures may decrease transmission system efficiency and could reduce available transmission capacity, according to DOE. Approximately 7 percent of generated power is lost in transmission and distribution, according to information publicly available on the EIA's website. As temperatures rise, the capacity of power lines to carry current decreases, according to DOE, as does the overall efficiency of the grid. Higher temperatures may also cause overhead lines to sag, posing fire and safety hazards. All of these factors can contribute to power outages at times of peak demand, according to USGCRP. In 2006, for example, electric power transformers failed in Missouri and New York, causing interruptions of the electric power supply in the midst of a widespread heat wave.

Broad, Systemic Factors Could Amplify Climate Change Impacts on Energy Infrastructure

Based on our previous work, as well as reports from USGCRP, NRC, and others, we identified several broad, systemic factors that could amplify the effects of climate change on energy infrastructure. These factors—which include changes in water availability, system interdependencies, increases in energy demand, and the compounding effects of multiple climate impacts—could have implications that extend throughout the energy sector and beyond.

Changes in Water Availability May Significantly Impact Energy Supply

As our series of reports on the energy-water nexus has shown, water and energy are inextricably linked and mutually dependent, with each affecting the other's availability.[48] Many aspects of energy production require the use of water to operate (see fig. 7). As discussed earlier in this review, fossil fuel and nuclear power plants—which accounted for about 90 percent of U.S. energy consumption in 2011—rely heavily on water for cooling purposes. As we reported in 2012, recently developed hydraulic fracturing methods also require significant amounts of water—3 to 5.6 million gallons of freshwater per well, according to our previous work on shale resources and development.[49] Increased evaporation rates or changes in snowpack may affect the volume and timing of water available for hydropower.

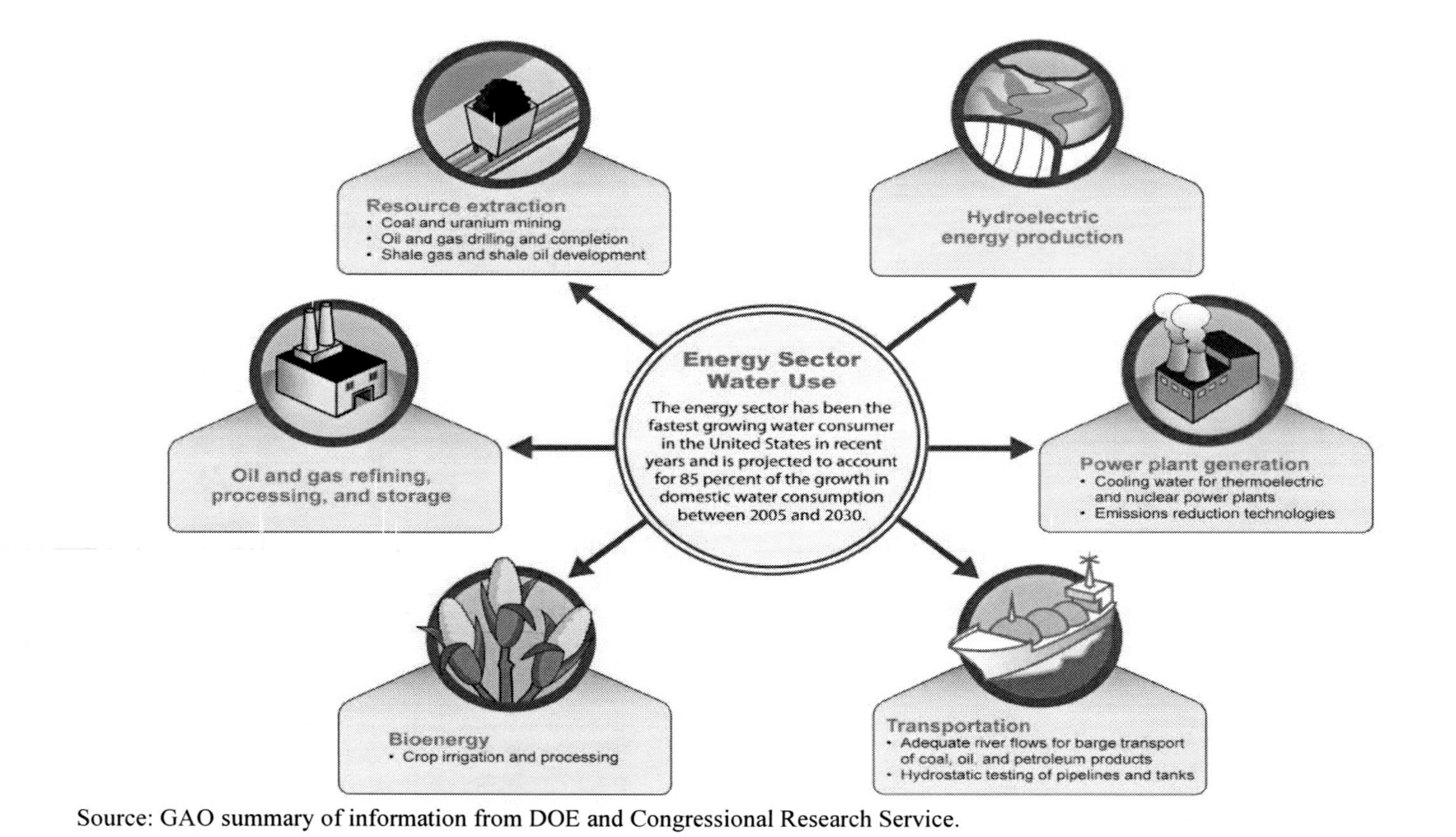

Source: GAO summary of information from DOE and Congressional Research Service.

Figure 7. Water Use by the U.S. Energy Sector.

Water is also required to mine and transport coal and uranium; to extract, produce, and refine oil and gas; and to support crops used in biofuel production, among other uses. According to the Congressional Research Service, the energy sector is the fastest growing water consumer in the United States and is projected to account for 85 percent of the growth in domestic water consumption between 2005 and 2030. This increase in water use associated with energy development is being driven, in part, by rising energy demand, increased development of domestic energy, and shifts to more water-intense energy sources and technologies.[50]

According to USGCRP and NOAA, increasing temperatures and shifting precipitation patterns are causing regional and seasonal changes to the water cycle—trends that present significant risks for the U.S. energy sector. More frequent and intense droughts, reduced summertime precipitation, and decreased streamflows are likely to adversely affect available water supply in some regions, particularly during the summer months.[51] Given the energy sector's dependence on water, these changes are likely to have wide-ranging impacts on the costs and methods for extracting, producing, and delivering fuels; the costs and methods used to produce electricity; the location of future infrastructure; and the types of technologies used. In recent years, a number of weather and climate events have served to illustrate some of the risks associated with water scarcity, as reported by DOE:

- In 2010, below-normal precipitation and streamflows in the Columbia River basin resulted in insufficient hydropower generation to fulfill load obligations for the Bonneville Power Administration, resulting in reported losses of approximately $233 million or 10 percent from the prior year;
- In 2007, a severe southeast drought reduced river flow in the Chattahoochee River by nearly 80 percent; reducing hydroelectric power in the Southeast by 45 percent;
- In 2012, drought and low river levels disrupted barge transportation of petroleum and coal along the Mississippi River.

USGCRP and DOE assessments further note that the energy sector's demand for water will increasingly compete with rising demand from the agricultural, industrial, and other sectors.

Energy Sector Interdependencies Can Amplify Impacts

The energy sector comprises a complex system of interdependent facilities and components, and damage to one part of the system can adversely affect infrastructure in other phases of the supply chain, according to DOE and USGCRP. Many different types of energy infrastructure—from pipelines to refineries—depend on electricity to function; as such, they may be unable to operate in a power outage, even if otherwise undamaged. Recent events associated with Hurricane Sandy illustrate these interdependencies—over 7,000 transformers and 15,200 poles were damaged, according to DOE, causing widespread power outages across 21 states. These outages affected a range of infrastructure dependent on electricity to function—for example, two New Jersey refineries were shut down, and a number of petroleum terminals and gas station fuel pumps were rendered inoperable. Because many components of the U.S. energy system—like coal, oil, and electricity— move from one area to another, extreme weather events affecting energy infrastructure in one region can lead to significant supply consequences elsewhere, according to USGCRP.

Interdependencies also link the energy sector to other sectors, such as transportation, agriculture, and communications. The energy sector requires railways, roads, and ports to transport resources such as coal, oil, and natural gas, for example; conversely, many modes of transportation rely on oil, gasoline, or electricity. Given these interdependencies, disruptions of services in one sector can lead to cascading disruptions in other sectors. Interrelationships between the transportation, fuel distribution, and electricity sectors were found to be major factors in Florida's recovery from past hurricanes, according to USGCRP.

Higher Temperatures Are Expected to Increase Energy Demand

Increases in temperature are expected to affect the cost, type, and amount of energy consumed in the United States, according to NRC and USGCRP assessments. Over the past four decades, the demand for cooling has risen and the demand for heating has declined (see fig. 8). As average temperatures rise and extreme weather events—such as heat waves—become more common, these trends are expected to continue, although specific impacts will vary by region and season.[52]

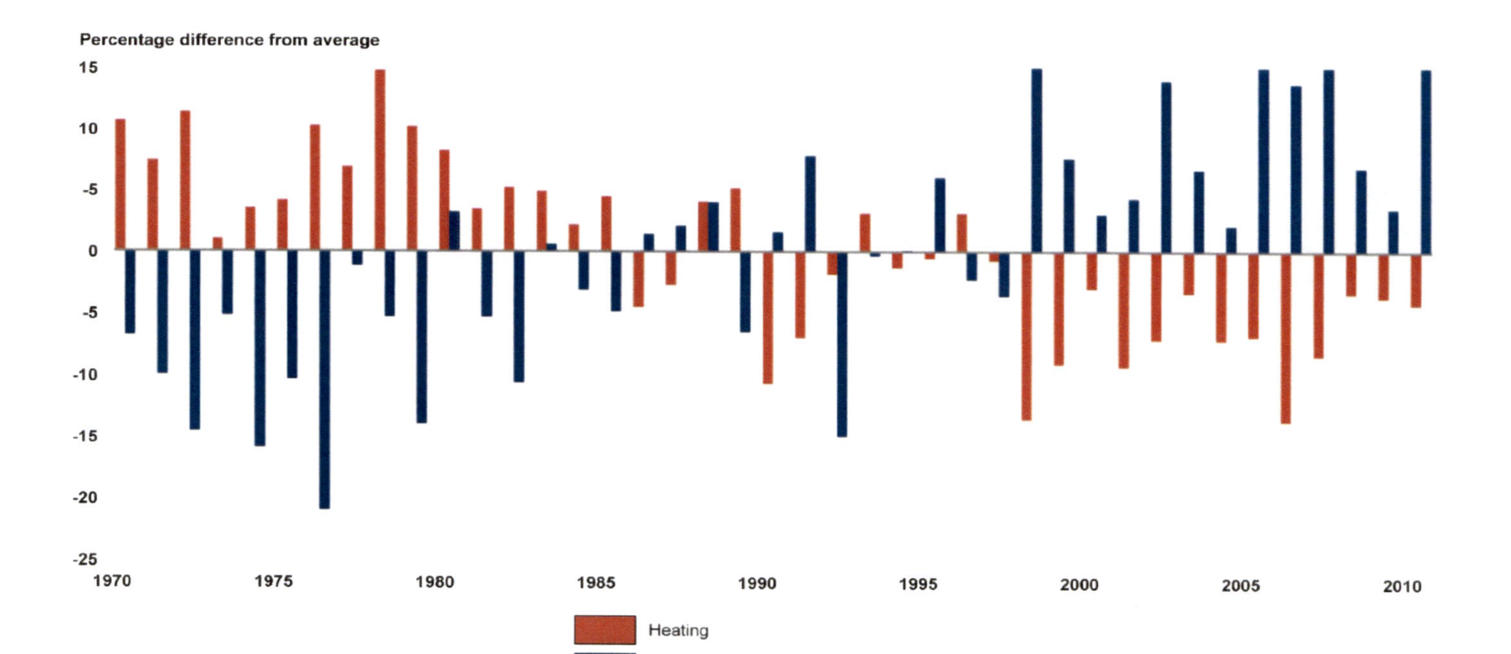

Source: United States Global Change Research Program 2013 Draft Climate Assessment.

Note: The amount of energy needed to cool (or warm) buildings is proportional to cooling (or heating) degree days. The figure shows increases in "cooling degree days" that result in increased air conditioning use, and decreases in "heating degree days," meaning less energy required to heat buildings in winter, compared with the average for 1970-2000.

Figure 8. Historical Increases in Cooling Demand and Decreases in Heating Demand.

Net electricity demand is projected to increase in every U.S. region, but particularly in southern states, since homes and businesses depend primarily on electricity for air conditioning. (In contrast, heating is provided by a combination of natural gas, heating oil, electricity, and renewable sources.) Increases in peak electricity demand caused by extreme high temperatures could potentially strain the capacity of existing electricity infrastructure in some regions, according to DOE.

In the summer heat wave of 2006, for example, some Midwest nuclear plants were forced to reduce output and several transformers failed, causing widespread electricity interruptions and making it difficult to access air conditioning.

Climate change-related increases in demand could also be exacerbated by a number of ongoing trends, such as population growth and increased building sizes.

Multiple Climate Impacts May Have Compounding Effects

According to DOE and IPCC, some climate change impacts are likely to interact with others, creating a compounding effect.[53] For example:

- Higher air and water temperatures may contribute to both an increase in electricity demand and a decrease in electricity supply.[54]
- The effects of sea level rise may be exacerbated by more severe storms and coastal erosion, causing flooding across a larger area. Storms can also damage natural features, such as wetlands, and manmade structures, such as sea walls, that help protect coastal infrastructure from sea level rise and storm surges.
- Both warmer temperatures and drought heighten the risk of flooding and wildfires, which—alone or in combination—could ultimately limit the amount of electricity that can be generated and transmitted during times of peak demand.

Compounding factors may be important for climate preparedness from both a local perspective as well as a regional or national perspective focused on overall system resilience, according to DOE.

ADAPTIVE MEASURES COULD REDUCE POTENTIAL CLIMATE CHANGE IMPACTS ON U.S. ENERGY INFRASTRUCTURE

Adaptive measures could reduce the potential for climate change to affect the energy infrastructure in the United States. As previously discussed, these measures vary across the energy supply chain, but they generally fall into two broad categories—hardening and resiliency. Industry decision makers we spoke with provided examples that illustrate some of the steps they have taken to integrate adaptive measures into their energy infrastructure, including investments that hardened their physical assets— such as elevating electrical substation control rooms to reduce potential flooding hazards— and improved the resiliency of portions of their energy supply system—such as purchasing backup power generators to restore electricity more quickly after a potential utility power outage.

Adaptive Measures Can Be Employed Across the Energy Supply Chain

While potential adaptive measures vary widely across the energy supply chain, they all generally focus on hardening—physical changes to make particular pieces of infrastructure less susceptible to storm-related damage—or improving resiliency— increasing the ability to recover quickly from damage to facilities' components or to any of the external systems on which they depend. For instance, hardening energy infrastructure across the supply chain is part of the energy industry's normal responsibilities and operating practices to ensure existing infrastructure is available to deliver energy to its customers under a range of weather conditions. According to industry representatives, industry chooses to make physical changes to its infrastructure to make it less likely to be damaged by extreme winds, flooding, or other weather events. Choices to harden infrastructure can require significant investment by industry, according to DOE's 2010 report on hardening and resiliency, such as building flood walls around refineries, elevating pumps used to transport fuels via pipelines, building power plants at higher elevations to minimize the risk of flooding, and replacing transmission and distribution poles with poles made of stronger materials to make them less susceptible to damage from high winds and storms.

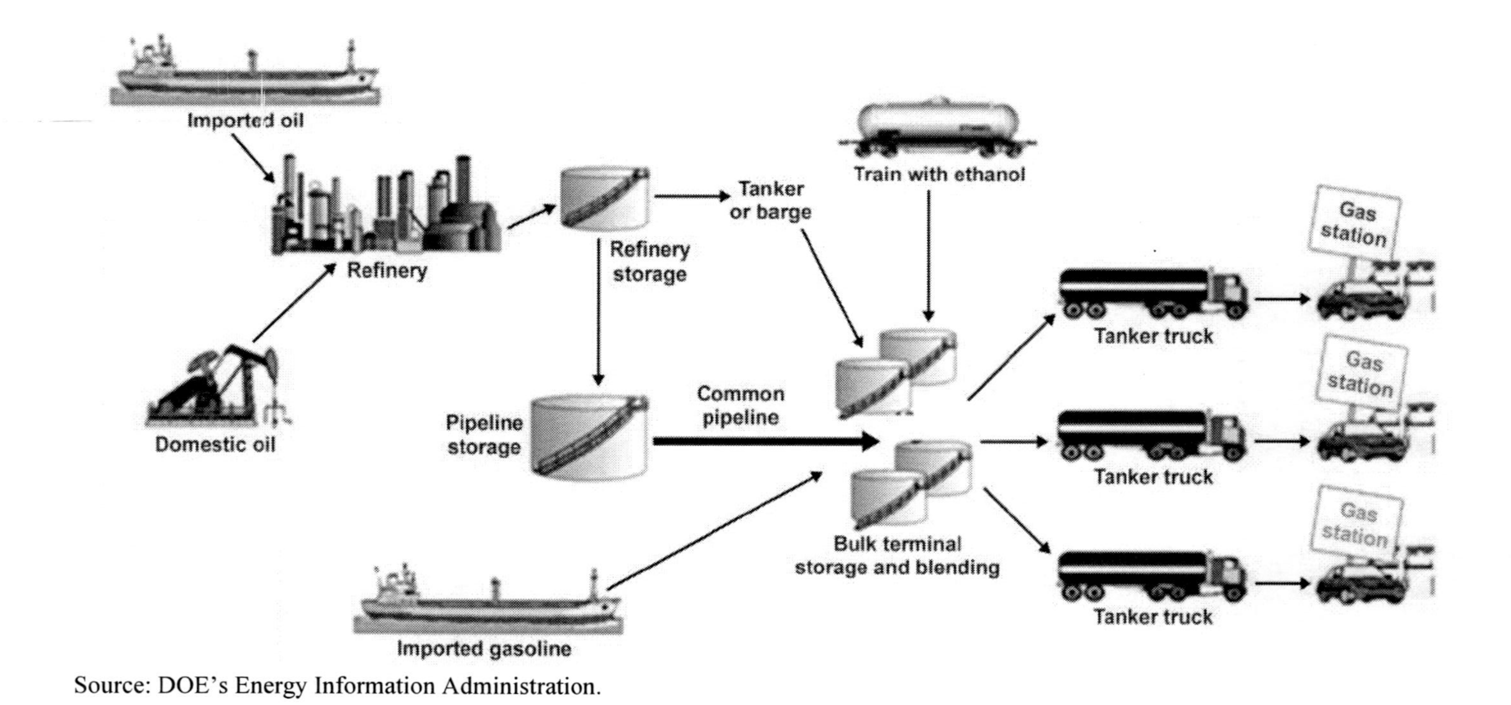

Source: DOE's Energy Information Administration.

Figure 9. The Gasoline Supply Chain.

In contrast to hardening measures that try to prevent damage, resiliency measures are focused on quickly recovering from damage to various parts of the energy supply chain, thereby enabling the system to continue to operate. Resiliency can take many forms and can be implemented by industry participants anywhere along the energy supply chain.[55] The following example—using one part of the energy supply chain, gasoline supplies—illustrates resiliency to potential events related to climate changes. In this illustration, if climate change resulted in rising sea levels that accentuate the damaging effects of tropical storms on the infrastructure for extracting, refining, transporting, or distributing oil, adaptation efforts in the various related parts of this infrastructure (see fig. 9) could help improve the overall resiliency in the gasoline supply chain. Specifically, at the beginning of the chain, adaptation could take the form of decreased extraction of oil from vulnerable offshore platforms supplanted by increased extraction from or use of less vulnerable onshore and foreign sources of oil.

Further down the supply chain, adaptation might involve decreased refining of oil from vulnerable refineries supplanted by increased refining of oil from less vulnerable refineries and additional imported gasoline from foreign refineries. Still further down the supply chain, if climate change rendered one mode of transport or distribution more vulnerable than others, adaptation might involve shipping or distributing less gasoline via the more vulnerable mode. Substitute sources of oil, refining, and transportation for the development and distribution of gasoline, therefore, represent ways in which industry can choose to adapt and limit disruptions to gasoline infrastructure and supply.

Examples of actual gasoline supply chain resiliency are demonstrated by actions taken during Hurricanes Katrina in 2005 and Sandy in 2012 as follows:

- In 2005, oil platforms were evacuated and damaged, as a result of Hurricane Katrina, virtually shutting down all oil production in the Gulf region. In response, the Administration approved loans of oil from the Strategic Petroleum Reserve to help refineries offset this short-term physical supply disruption at the beginning of the supply chain, thereby, helping to moderate the impact the production shutdown had on U.S. crude oil supplies.[56]
- A more recent example, following Hurricane Sandy, illustrates how such alternatives can help increase resiliency at the distribution stage of the supply chain. In 2012, this storm damaged petroleum terminals used to store and distribute gasoline in the New York Harbor (NYH)

area, thus disrupting the normal supply chain. However, according to DOE's EIA report on the summary of impacts on petroleum supplies following Hurricane Sandy, "...areas normally served by the NYH terminals were also receiving some supplies through more distant terminals as industry pursued workarounds to meet customer needs to the best of their ability." [57] Thus, while a significant disruption in the overall ability to move gasoline through the NYH area occurred as a result of the storm, other terminals outside the affected area helped to ameliorate some of the supply loss.

Ultimately, how much adaptation will take place and in what form—hardening and increasing resilience, such as choosing substitute actions as described above—will depend on how the costs of adaptation compare with the expected costs of taking no action.[58]

Industry Decision Makers Have Taken Steps to Integrate Adaptive Measures into Energy Infrastructure

Industry decision makers have taken steps to integrate adaptive measures into energy infrastructure planning and investments using varying approaches as illustrated by the following examples. In three of the examples we selected, companies implemented a company-wide approach and incorporated several adaptive measures into overall energy infrastructure planning and investments. In another example we selected, a company increased resiliency by purchasing and prepositioning mobile generators to run key facilities and pumping stations along oil pipelines in the event of power outages.

Entergy

Entergy Corporation generates, transmits, and distributes electric power in the Southeast. According to Entergy representatives, its transmission and distribution infrastructure along the coast of the Gulf of Mexico is vulnerable to extreme weather events and storms, storm surge caused by hurricanes, and sea level rise associated with land subsidence. Following Hurricanes Katrina and Rita in 2005, Entergy experienced unprecedented damage, leading to power outages for roughly 800,000 customers in Louisiana. The company faced widespread damage to transmission and distribution systems, flooded substations, and power plants resulting in shutdowns. Figure 10 shows an example of damaged transmission power lines as a result of Hurricane Rita.

Entergy's New Orleans subsidiary— Entergy New Orleans (ENO)—filed for bankruptcy after this major damage to its infrastructure and the declining revenues due to the drastic reduction to its customer base as residents left the city.

Source: Energy.

Figure 10. Fallen Transmission Lines after Hurricane Rita.

Driven by a lack of useful information on which to base planning for infrastructure protection against future storms, Entergy representatives told us that they partnered with the America's Wetland Foundation (AWF) and commissioned a study in 2010 identifying the company's most critical and vulnerable assets in the Gulf.[59] The study also highlighted adaptation strategies that have low investment requirements, high reduction of expected losses—regardless of climate change impacts—and additional benefits, such as coastal wetlands restoration. For example, Entergy representatives told us that the study identified a number of potential hardening and resiliency measures, such as replacing wooden transmission and distribution poles with steel or concrete, strengthening distribution poles, building levees and berms around oil refineries, elevating substations in flooding areas, and managing vegetation along electricity lines. The study estimated potential losses of $350 billion along the Gulf Coast by 2030 due to rising sea level and loss of coastline. It also identified $120 billion in potential investments and concluded that supporting a range of adaptive actions to reduce the potential weather and climate-related risks, and identifying barriers to increasing industry resilience, are important elements of a coordinated response.

Entergy representatives told us by taking a company-wide approach to identify infrastructure vulnerable to climate-related risks, they implemented several adaptive measures highlighted in the study, such as replacing wooden transmission poles with steel, strengthening distribution poles with support wires, and elevating sensitive electronic equipment in select substations. In response to more recent storms, such as Hurricane Isaac, Entergy representatives told us that the implementation of these adaptive measures has paid off. They have experienced less infrastructure damage and have restored power to their customers more quickly than in previous storms.[60]

Pacific Gas and Electric Company

Pacific Gas and Electric Company (PG&E) provides natural gas and electric power to 15 million people in northern and central California. PG&E continues to implement a company-wide approach to incorporate climate-related risks as part of its business planning and investments. In 2008, PG&E convened a science team--specializing in meteorology, biology, and hydrology— to evaluate global climate-related risks, assess climate change modeling, and identify best adaptation practices for the company's assets. PG& E officials told us that risks and recommendations developed by the science team are used to develop adaptation strategies for infrastructure potentially impacted by weather and climate-related risks such as sea level rise, increased air temperatures, and changes in precipitation patterns.

For example, PG&E representatives told us that some equipment in the company's substations is vulnerable to increased temperatures in the state. Therefore, PG&E worked with the equipment manufacturer's engineers on best operating practices for their substations at higher temperatures. Additionally, PG&E has major transmission and distribution lines in the San Francisco Bay Area that are potentially susceptible to sea level rise. Therefore, the company has strengthened electric transmission structures in the southern Bay Area and is collaborating with state and federal agencies on bay habitat restoration that will help increase utility resiliency to tidal action, according to PG&E officials. See figure 11 for a picture of Gateway Generating Station.

Source: PG&E 2011 Corporate Responsibility and Sustainability Report.

Figure 11. Gateway Generating Station.

As a result of limited water availability and other factors, PG&E implemented dry cooling technology at two of its natural gas fueled generating stations, Gateway Generating Station in Antioch, California (2007), and Colusa Generating Station in Maxwell, California (2010). Although reducing plant efficiency under some conditions, dry cooling technology uses 97 percent less water and produces 96 percent less discharge than a conventional water cooling system, which helps the company significantly reduce the use of water for cooling purposes. PG&E representatives cited incorporation of a less water intensive technology as having increased the plants' resiliency to potentially decreased water availability.

Scientists also predict that climate change will result in significant reductions in snowpack in parts of the Sierra Nevada Mountains, potentially impacting PG&E's hydroelectric system. PG&E's adaptation strategies include developing new modeling tools for forecasting runoff, maintaining higher winter carryover reservoir storage levels, reducing conveyance flows in canals and flumes during winter storms as more precipitation falls as rain, and reducing discretionary reservoir water releases, according to PG&E officials.

Colonial Pipeline

Colonial Pipeline owns and operates a 5,500-mile network of pipelines running from Houston, Texas, to NYH. [61] These pipelines transport a daily average of 100 million gallons of refined petroleum products such as gasoline, diesel fuel, home heating oil, fuels for commercial aviation and for the U.S. military—accounting for about 15 percent of the fuel supplied in the United States and almost 65 percent of fuel supplied in the Southeast. In general, such pipeline systems are susceptible to disruption from severe weather events, primarily because they require significant amounts of electric power to operate computer systems, generators, and pumps. Disruptions to this power can reduce or halt the transport of refined products in the pipeline system.

Colonial representatives told us that to enhance resiliency after the 2005 hurricane season along the Gulf Coast, Colonial purchased 12 large mobile generators (i.e., Gensets) and seven transformers to help it recover more quickly from power losses due to severe weather events (see fig. 12). According to these representatives, this equipment allows the pipeline company to run key pumping stations anywhere along the pipeline to minimize disruptions when electric power is unavailable due to severe weather or other events.

Source: DOE.

Figure 12. Colonial's Portable Generators.

In addition, after Hurricanes Gustav and Ike in 2008, Colonial implemented a number of resiliency measures, such as monitoring storm paths to preposition generators where power would most likely be lost. Company representatives told us they also used Colonial's Control Center in Atlanta, Georgia, to communicate with fellow employees about potential areas where the pipeline might experience power outages. The company followed similar efforts in preparation for Hurricane Sandy in 2012. For example, Colonial representatives moved one-half of the company's new mobile generators from Mississippi to Linden, New Jersey, prior to Hurricane Sandy making landfall.[62] After the storm, while Colonial's pipeline system remained undamaged, electrical power was down, but company representatives told us they successfully used the mobile generators to restore power to the pipeline, resulting in relatively few disruptions for oil transportation along the system. Figure 12 shows an example of portable generators used to transport oil during electrical outage.

Florida Power and Light

Florida Power and Light (FPL)—the largest electric utility in Florida—generates and distributes electricity to approximately 4.5 million customers. FPL representatives told us that one of the company's nuclear power plants, Turkey Point in Homestead, the largest generating station in Florida, is potentially vulnerable to extreme weather events and storms, storm surge caused by hurricanes, and sea level rise. In response, FPL has implemented a company-wide approach to incorporate climate-related risks into their infrastructure planning and investments. FPL representatives told us they have a vested interest in hardening existing and new infrastructure to withstand climate change impacts given the substantial capital expense that the company invests in this infrastructure. For example, the current power plant, which was built in the 1960s, is elevated 18 feet above sea level to protect against flooding and extreme storm surges. According to company representatives, all equipment and components important to nuclear safety are protected to about 20 feet above sea level and protected from waves to about 22 feet above sea level on the side facing Biscayne Bay.

In June 2009, FPL submitted an application to the Nuclear Regulatory Commission to evaluate an option for constructing and operating two new nuclear reactor units— Units 6 and 7—at the existing Turkey Point site. As part of its reactor licensing process, the Nuclear Regulatory Commission requires licensees to assess and if necessary, take measures to mitigate the impacts of the natural hazards their reactors might face.[63] As part of this permitting and licensing process, FPL's natural hazard assessment for Units 6 and 7 incorporated potential sea level rise over the next 100 years. According to company representatives, their hazard assessments for Units 6 and 7 used assumptions that are at least 20 percent more conservative than those used in the 1960s. For example, based on an extrapolation of historical weather data, FPL calculated that potential sea level rise over the next 100 years would be about 9 inches. The company rounded up the estimate to 1 foot of sea level rise to account for the uncertainties of potential climate change. Additionally, in March 2013, FPL representatives submitted a reevaluation of flooding hazards for Units 3 and 4 to the Nuclear Regulatory Commission that also incorporated projected sea level rise over the next 20 years when the existing reactors' license expires. FPL's reevaluation calls for it to use the latest available information and methodologies to analyze site-specific hazards, including stream and river flooding, hurricane storm surges, tsunamis, and dam failures. This reevaluation will determine whether the hazard exceeds the facility's flooding design basis so the Commission can assess the safety of the existing reactors at the Turkey Point site in light of more recent information. Figure 13 is a proposed illustration of Turkey Point's Nuclear Units 6 and 7.

Source: Florida Power and Light.

Figure 13. Proposed Turkey Point Nuclear Units 6 and 7.

Federal Role in Directly Adapting Energy Infrastructure Is Limited, but Selected Federal Entities Can Play an Important Supporting Role in Decision Making and Are Initiating Actions toward Adaptation

The federal government has a limited role in directly adapting energy infrastructure to the potential impacts of climate change, but selected federal entities can play important supporting roles that influence private companies' investment decisions and are taking steps to begin adaptation efforts within their respective missions.

Federal Influence on Energy Infrastructure Adaptation Decisions Generally Falls into Four Areas

Energy infrastructure adaptation is primarily accomplished through planning and investment decisions made by private companies that own the infrastructure;[64] nevertheless, the federal government can influence private sector investment decisions through: (1) providing information, (2) regulatory oversight, (3) technology research and development, and (4) market incentives and disincentives.

Providing Information

The federal government plays an important role in providing information to promote climate resilience. As we reported in our 2013 High Risk report, federal efforts on climate change are beginning to shift their focus to adaptation and providing information to state and local decision makers so they can make more informed decisions about the fiscal exposure posed by potential climate changes.[65] Several federal agencies play a role in providing this information, including NOAA, DOE, U.S. Geological Survey, and USGCRP, as follows:

- NOAA develops and shares weather and climate-related information with government officials and private industry. For example, NOAA officials told us that through the National Weather Service (NWS) it produces weather forecasts for local areas out to seven days and probabilistic climate outlooks from 6 days out to a year. According to officials, NWS also monitors and assesses the state of the climate and provides information on longer term climate cycles such as the El Nino Southern Oscillation cycle.[66] Some industry decision makers told us that they use these data when making infrastructure planning and implementation decisions.

- DOE is working to make more climate change information available for decision makers at the federal, state, and local levels. For example, DOE officials stated that the Office of Science supports research reviewing available climate models and scientific projections, and it looks at local climate models in order to build in more locally detailed information to be useful and available to decision makers on a timely basis. In addition, DOE officials stated that the Office of Science supports research analyzing extreme weather events, including floods and droughts, and how these events impact regions.
- USGS provides fundamental scientific information, tools, and techniques that land, water, wildlife, and cultural resource managers and other decision makers can apply to anticipate, monitor, and adapt to climate change impacts. Also, according to the USGCRP 2012-2021 strategic plan,[67] USGS scientists have worked in collaboration with other USGCRP agencies to meet the needs of policymakers and resource managers for scientifically valid state-of-the-science information and predictive understanding of global change and its effects.
- USGCRP's strategic objectives for 2012-2021 include improving the deployment and accessibility of science to inform adaptation decisions. To this end, USGCRP states in its strategic plan that its member agencies will work with state, local, and tribal governments, and other federal agencies to build the capabilities for engagement and support needed by all decision makers, especially in key areas of vulnerability. Additionally, in coordination with USGCRP, the President's Climate Action Plan stated for the first time, the 2014 National Climate Assessment will focus not only on dissemination of scientific information but also on translating scientific insights into practical, useable knowledge that can help decision makers anticipate and prepare for specific climate-change impacts.[68]

Regulatory Oversight

The federal government can also play a supporting role in energy infrastructure investment decisions through its regulatory oversight role. EPA,

FERC, the Nuclear Regulatory Commission, as well as NERC, regulate energy infrastructure by promulgating and enforcing emissions, reliability, and safety standards as follows:

- EPA issues environmental regulations that can have implications for electricity generation facilities, petroleum refineries, oil and gas extraction facilities, natural gas pipelines and hydrocarbon storage wells. EPA's influence over these types of energy infrastructure comes when owners must make technological changes to the infrastructure in order to comply with EPA's regulatory requirements. For example, in July 2012, we reported that EPA regulations will require some electricity generation facilities to install additional emissions controls. In some cases, electricity generation facilities may convert from coal to natural gas or even shut down rather than install the emissions controls necessary to comply with the regulations.[69]
- FERC's influence on energy infrastructure comes primarily through its review and authorization process for specific projects. Through the review and authorization process, FERC can change the siting and design of a facility, require environmental mitigation measures and, for hydroelectric and natural gas infrastructure, can impose safety requirements. FERC's activities have implications for several types of energy infrastructure, including hydropower plants, interstate natural gas pipelines and storage facilities, liquefied natural gas facilities, and interstate oil pipelines.
- The Nuclear Regulatory Commission, as part of its reactor licensing process, evaluates nuclear power plant specifications, including requirements for flood protection. Determining how high a plant should be built to be safe from flooding is critical for U.S. nuclear power plants located on the coast given the lack of scientific consensus on the actual rate of sea level rise.
- NERC, through an industry-based consensus development process, establishes and enforces reliability standards for the bulk power system. These standards can result in energy infrastructure owners making technological changes to their infrastructure in order to ensure that the grid operates reliably.[70] NERC also investigates and analyzes the causes of significant power system disturbances in order to help prevent future events.

Technology Research and Development

DOE also plays an important role in the research and development of new technologies to support the energy industry. In general, DOE conducts and funds a wide array of research and development programs aimed at both understanding the impact of climate change on energy production and developing new technologies to improve resilience to climate change. DOE also conducts assessments of climate change on electric grid stability, water availability for energy production, and site selection of the next generation of renewable energy infrastructure. For example, DOE officials stated that its Office of Energy Efficiency and Renewable Energy is looking at how to make biofuels and biomass less dependent on water, and DOE's report on energy sector vulnerabilities identified this area of research as a technology opportunity where combined public and private efforts to improve the resilience of the energy sector should increase. The report also cites several other specific examples of technological options to improve climate resilience, including: enhanced restoration technologies and practices to maintain or expand regional wetlands and other environmental buffer zones; increased power plant efficiency through integration of technologies with higher thermal efficiencies than conventional coal-fired boilers; and improved water reservoir management and turbine efficiency for more efficient hydropower generation.

Market Incentives and Disincentives

The federal government also provides a range of market incentives and disincentives that can encourage or discourage industry from implementing energy infrastructure adaptation measures. Incentives to incorporate adaptive efforts include tax incentives, direct expenditures and other support, such as production tax credits for renewable technologies. One example of direct expenditure is DOE's Smart Grid Investment Grant Program. The program is structured as a public–private partnership to accelerate investments in grid modernization. DOE reports that $3.4 billion dollars in federal Recovery Act funds were matched with private sector resources—bringing the total investment to about $7.8 billion. These funds were used to support 99 projects that are now deploying smart grid technologies in almost every state.[71] Smart Grid technology incorporates the usage of smart meters that have outage notification capabilities that make it possible for utilities to know when customers lose power and to pinpoint outage locations more precisely. Smart meters also indicate when power has been restored.

At the same time, some federal programs may discourage adaptation efforts that, in turn, can impact energy infrastructure. For example, according to studies we reviewed, to the extent that federal insurance programs—such as the National Flood Insurance Program (NFIP)[72] — have set premiums that are not risk-based, they can discourage individuals from engaging in adaptation that might otherwise have occurred. Federal Emergency Management Agency (FEMA) officials estimate that currently about 20 percent of NFIP policyholders pay less than full-risk premiums, which FEMA refers to as subsidized premiums. FEMA officials also estimate that, on average, policyholders with subsidized premiums pay only about 45 percent to 50 percent of the full-risk premium. In such instances, adaptation may have been discouraged because such premiums have lowered the incentive for individuals to adapt by subsidizing investments that do not take into account the potential impacts of climate change. In general, as population increases along U.S. coastal areas more prone to damage from extreme weather, sea level rise, and high winds, infrastructure is built to provide and extend essential services such as energy and water. As a result, this infrastructure maybe more vulnerable to changing weather and climate conditions than it might have been had it been located further from the coast.

Selected Federal Entities That Can Influence Energy Infrastructure Adaptation Are Beginning to Take Steps to Address Climate Change Risks

Selected federal entities that can influence energy infrastructure adaptation decisions are beginning to address climate change risks through project-specific activities, as well as through broader agency-wide assessments and interagency cooperation. Both the project-specific activities and the broader agency-wide assessments could impact energy infrastructure adaptation, but it is too early in the agencies' assessment process to understand how, if it all, the assessments could influence energy infrastructure decision makers.

Selected Federal Entities Address Climate Change Risks on a Project-Specific Basis

DOE, EPA, FERC, the Nuclear Regulatory Commission, as well as NERC, are beginning to incorporate consideration of climate change risks on a project-specific basis. Examples include the following:

- DOE has a long history of conducting fundamental energy science and energy technology research and development, and climate change is an ongoing part of DOE research, modeling and policy development. DOE program offices support a range of research and development activities related to climate change, according to DOE officials. For example, DOE's Office of Fossil Energy and the National Energy Technology Laboratory (NETL) are developing advanced water management technologies applicable to fossil and other power plants in three specific areas: nontraditional sources of process and cooling water to demonstrate the effectiveness of utilizing lower quality water for power plant needs[73] ; innovative research to explore advanced technologies for the recovery and use of water from power plants; and advanced cooling technology research that examines wet, dry, and hybrid cooling technologies. This research, like other NETL activities that move innovations from the lab to the marketplace, can help advance the adaptive efforts that private companies are making to incorporate less water-intensive technology.

- EPA's Office of Water implements the National Water Program (NWP), which, according to EPA officials, monitors changes in power generation across the United States and the impacts of those changes on water resources. The NWP has developed two climate change strategies, the first in 2008 and the second in 2012. The most recent 2012 strategy recognizes that water and energy are intimately connected, and it puts forth the goal of using a systems approach to reduce the demand for both water and energy. The systems approach is one in which water, energy, and transportation infrastructure planning is integrated in order to increase efficiencies for all three sectors.
- FERC officials told us that while they consider climate models as a part of their review, FERC makes decisions on a project-by-project basis, and general modeling information alone is not sufficient. For example, during the review and authorization processes for liquefied natural gas terminals, FERC evaluates whether terminal operators have accounted for potential hurricane and flooding impacts. Similarly, for hydropower project review and authorization, FERC noted that it looks at historical hydrological data as part of its analysis of project operation, which often includes monitoring and a provision that allows FERC to alter license requirements should environmental conditions change in the future. For example, if water levels change, officials can require project-specific adaptation changes that account for regional conditions, such as a drought in the Southeast. If drought

conditions continue and less rainfall is expected, they may also suggest adaptive measures for a particular period of time.

- The Nuclear Regulatory Commission uses the reactor licensing process to review whether individual projects have adequate protection against hurricanes, flooding, and other natural phenomena. The Nuclear Regulatory Commission has revised its guidance for hurricane wind speed protection at nuclear power plants. It has also required the operating fleet of nuclear power plants to complete reevaluations of certain hazards (i.e., seismic and flood hazards) using updated hazard information and present-day guidance and methodologies. Further, Nuclear Regulatory Commission officials noted that, if natural phenomena are shown to have the potential to cause a plant to exceed its safety parameters, the plant must correct the issue or be subject to enforcement actions.[74] These officials told us that they have access to the same data that insurance companies use in their climate modeling as well as other information to evaluate such actions on plant safety and operations.
- NERC uses historical weather data to help it assess the reliability of the electrical grid, including changes that might result from climate change. NERC looks at weather patterns as part of its summer, winter, and 10-year reliability assessments that provide an overview of projected electricity demand growth, as well as other information on generation and transmission additions. Additionally, NERC officials told us that NERC's annual 10-year reliability assessments provide an independent view of the reliability of the electrical grid, identifying trends, emerging issues, and potential concerns. NERC's projections are based on a bottom-up approach, collecting data and perspectives from grid operators, electric utilities, and other users and owners, of the electrical grid.

DOE and EPA Are Initiating Plans to Address Climate Change Risks Using an Agency-wide Approach

DOE and EPA are also beginning to incorporate consideration of climate change risks on an agency-wide basis, via agency climate change adaptation plans and the Interagency Climate Change Task Force. Executive Order 13514 on Federal Leadership in Environmental, Energy, and Economic Performance required federal agencies to develop strategic sustainability performance plans, which include climate change adaptation plans.[75] The implementing instructions for developing the climate change adaptation plans stated that, through adaptation planning, each agency will identify aspects of climate change that are likely to impact the agency's ability to achieve its mission and sustain its operations and respond strategically. According to the implementing instructions, integration of climate change adaptation planning into the operations, policies, and programs of the federal government will ensure that resources are invested wisely and that federal services and operations remain effective in current and future climate conditions. Federal agencies, including DOE and EPA, publicly released their first climate change adaptation plans in February 2013, as part of their annually updated strategic sustainability performance plans.[76]

These initial adaptation plans provide a high-level vulnerability analysis of the impact of climate change on the agencies' mission, operations, and programs but do not address specific actions the agencies will take or how those actions could influence decision makers. Within DOE's adaptation plan, the actionable items for integrating climate change resilience focus on updating departmental planning documents to include climate adaptation planning considerations. Thus, the details of how DOE will take action on climate change across its programs is not yet known, although DOE will continue to update and modify DOE's adaptation plan as the understanding of climate change improves. Similarly, EPA's initial adaptation plan states that EPA expects to improve its understanding of how to integrate climate change adaptation into its programs, policies, rules, and operations over time. However, EPA officials told us that they have not yet determined which rulemaking processes this will include.

In addition to developing climate change adaptation plans, DOE and EPA participate in the Interagency Climate Change Adaptation Task Force. Executive Order 13514 called for federal agencies to participate actively in the already existing Interagency Climate Change Adaptation Task Force.[77] According to agency officials, both DOE and EPA have been active members of the task force. For example, DOE's Office of Energy Policy and Systems Analysis and Office of Science participate in task force working groups on water availability and climate science. According to DOE officials, much of the task force's time is currently spent on sharing best practices and lessons learned about the implementation of adaptation efforts among agencies. Additionally, EPA officials told us that the agency leads a community of practice within the task force to bring federal agencies together to work on common issues and share lessons learned relating to climate change adaptation. As part of those meetings, task force members discuss available adaptation strategies and associated costs, as well as the costs of inaction and ways to finance adaptive measures. The task force's most recent progress report, released in October 2011, reported that the federal government was working to improve the accessibility and utility of climate information and tools (e.g., climate models and early-warning systems) to meet the needs of decision makers. As the task force continues its work in this area, more information on how these efforts impact energy infrastructure decision makers may become available. The task force's next update on federal adaptation progress is due in March 2014.

CONCLUDING OBSERVATIONS

A wide range of studies and years of industry experience have clearly demonstrated that U.S. energy infrastructure is at risk for damage and disruptions to service due to severe weather events. The damage from such events can impose large costs on the energy industry, as well as impact the economies of local communities and the nation. According to best available science, energy experts, as well as industry officials with whom we spoke, climate change could increase these risks unless steps are taken to adapt to expected changes. While uncertainty exists about the exact nature, magnitude, and timing of climate change, the responsibility for adapting energy infrastructure remains principally under the domain of the private sector. In this context, industry has and will continue to face choices about how best to respond to these risks.

At the same time, the federal government is just beginning to engage in more coordinated, multiagency efforts to better understand how climate change might impact federal facilities and their mission goals that intersect with the energy industry. As noted in our High Risk work, federal efforts related to infrastructure are beginning to focus on ways to help state and local governments make more informed decisions to adapt to climate change. Nascent federal efforts related to the energy sector may, in a similar way, provide an opportunity for agencies to consider how they could best inform private sector choices to adapt to climate change.

AGENCY COMMENTS AND OUR EVALUATIONS

We provided a draft of this report for review and comment to the Department of Energy (DOE), Environmental Protection Agency (EPA), Federal Energy Regulatory Commission (FERC), and the Nuclear Regulatory Commission. The Nuclear Regulatory Commission provided a general comment about the report which is summarized below. The Nuclear Regulatory Commission also provided technical and clarifying comments as did DOE and EPA which we incorporated as appropriate. FERC indicated that it had no comments on the report.

In general, the Nuclear Regulatory Commission clarified that its regulations do not directly address the impacts of climate change on nuclear power plants but that it requires these plants to be protected against the effects of certain natural phenomena, such as flooding and high winds. In as much as climate change affects those natural phenomena on a site-specific basis, the Commission said that climate change effects are considered. In response to this comment, we added language to the report to clarify that the Commission believes climate changes is taken into consideration as part of its review of natural phenomenon.

Frank Rusco
Director,
Natural Resources and Environment

APPENDIX I: OBJECTIVES, SCOPE, AND METHODOLOGY

This report examines: (1) what is known about the potential impacts of climate change to U.S. energy infrastructure; (2) measures that can reduce climate-related risks and adapt the energy infrastructure to climate change; and (3) the role of the federal government in adapting energy infrastructure to the potential impacts of climate change and steps selected federal entities have taken toward adaptation.

To address all three objectives, we reviewed relevant studies and government reports, including previous GAO reports. To identify relevant studies and reports, we conducted a literature review with the assistance of a technical librarian. We searched various databases, such as ProQuest, PolicyFile, and Academic OneFile and focused on peer reviewed journals, government reports, trade and industry journals, and publications from research organizations, advocacy groups, and think tanks from 2003 to 2013. To identify knowledgeable stakeholders, we reviewed our prior climate change work and relevant outside reports to identify individuals with specific knowledge of climate change adaptation and energy infrastructure. We then interviewed academics and knowledgeable professional association members from the American Gas Association, America's Wetland Foundation, the Center for Clean Air Policy, the Center for Climate and Energy Solutions, Ceres, the Environmental and Energy Study Institute, Georgia Institute of Technology, Louisiana State University, the National Association of Regulatory Utility Commissioners, the National Association of State Energy Officials, the National Rural Electric Cooperative Association, and the Nature Conservancy.

To examine what is known about the impacts of climate change on U.S. energy infrastructure, we reviewed climate change impact assessments from the National Research Council (NRC), the U.S. Global Change Research Program (USGCRP), and relevant federal agencies. We identified these assessments using government and National Academies websites and prior GAO reports on climate change. We then evaluated whether the assessments fit within the scope of our work and contributed to the objectives of this report. For relevant assessments, we used in-house scientific expertise to analyze the soundness of the methodological approaches they utilized, and we determined them to be sufficiently sound for our purposes. Relevant assessments are cited throughout this report, but the key assessments for this objective included the following:

- NRC, America's Climate Choices: Panel on Adapting to the Impacts of Climate Change, Adapting to the Impacts of Climate Change (Washington, D.C.: 2010).
- V. Bhatt, J. Eckmann, W. C. Horak, and T. J. Wilbanks, Possible Indirect Effects on Energy Production and Distribution in the United States in Effects of Climate Change on Energy Production and Use in the United States. A Report by the U.S. Climate Change Science Program and the subcommittee on Global Change Research (Washington, D.C..: 2007).
- Thomas R. Karl, Jerry M. Melillo, and Thomas C. Peterson, eds., Global Climate Change Impacts in the United States (New York, NY: Cambridge University Press, 2009).

- USGCRP, Draft Third National Climate Assessment Report, Chapter 4 – Energy Supply and Use (January 2013).
- Department of Energy, Infrastructure Security and Energy Restoration, Office of Electricity Delivery and Energy Reliability, Hardening and Resiliency U.S. Energy Industry Response to Recent Hurricane Seasons (August 16, 2010).
- Department of Energy, U.S. Energy Sector Vulnerabilities to Climate Change and Extreme Weather, DOE/PI-0013 (July 2013).

Using those assessments, we examined potential impacts to the following infrastructure categories, representing four main stages of the energy supply chain: (1) resource extraction and processing infrastructure; (2) fuel transportation and storage infrastructure; (3) electricity generation infrastructure; and (4) electricity transmission and distribution infrastructure. We also examined broad, systemic factors that may amplify climate change impacts to energy infrastructure.

To identify and examine measures to reduce climate-related risks and adapt energy infrastructure to climate change, we analyzed relevant studies and government reports and interviewed knowledgeable stakeholders. We also identified and selected a nonprobability sample of four energy companies where the companies were taking steps to adapt their energy infrastructure to the potential impacts of climate change: Colonial Pipeline Company, Entergy Corporation, Florida Power and Light Company, and Pacific Gas and Electric Company. To select our sample, we reviewed the relevant literature and interviewed knowledgeable stakeholders in order to compile a list of adaptive measures that had been undertaken by energy companies across the country. We divided our initial list of approximately 20 companies that had taken adaptive measures into categories based on geographic location and the type of climate-related risk that the adaptive measure addressed. We then selected companies that represented a range of geographic locations—California, Louisiana, Florida, and the Mid-Atlantic— and a range of climate-related risks—decreased water availability, increased frequency and intensity of storms, and increased precipitation. Our sample selection also reflects infrastructure used in three of the four stages of the energy supply chain: petroleum pipelines; hydropower, fossil fuel and nuclear power plants; and transmission and distribution infrastructure. Because this was a nonprobability sample, findings from our examples cannot be generalized to all U.S. energy infrastructure; instead, they provide illustrative information about energy companies that have undertaken adaptation measures.

To examine the role of the federal government in adapting energy infrastructure to the potential impacts of climate change including, what steps selected federal entities have taken towards adaptation, we identified federal entities with key responsibilities related to energy infrastructure. We began by reviewing relevant literature and interviewing knowledgeable stakeholder groups. We also interviewed officials from: Department of Energy (DOE), Department of the Interior (DOI), Department of Homeland Security (DHS), Department of Transportation (DOT), Environmental Protection Agency (EPA), Federal Emergency Management Agency (FEMA), Federal Energy Regulatory Commission (FERC), National Oceanic and Atmospheric Administration (NOAA), the Nuclear Regulatory Commission, as well as the North American Electric Reliability Corporation (NERC).[1] Based on the information gathered from the literature and interviews, we compiled an initial list of 15 federal entities that had a connection to energy infrastructure and then narrowed that list to the five that have the most direct influence on energy infrastructure adaptation decisions: DOE, EPA, FERC, the Nuclear Regulatory Commission, and NERC.

Appendix II contains one-page summaries of the federal government's role in energy infrastructure for DOE, EPA, FERC, and the Nuclear Regulatory Commission, as well as NERC. To develop the summaries, we reviewed and synthesized publically available information about the entities from their websites, prior GAO reports, and other reports describing the federal government's activities related to energy, as well as interviews with officials from the federal entities.

We conducted this performance audit from July 2012 to January 2014 in accordance with generally accepted government auditing standards. Those standards require that we plan and perform the audit to obtain sufficient, appropriate evidence to provide a reasonable basis for our findings and conclusions based on our audit objectives. We believe that the evidence obtained provides a reasonable basis for our findings and conclusions based on our audit objectives.

APPENDIX II: SUMMARIES OF SELECTED FEDERAL ROLES IN ENERGY INFRASTRUCTURE

Agency	**Department of Energy (DOE)**
Mission	Ensures America's security and prosperity by addressing its energy, environmental, and nuclear challenges through transformative science and technology solutions
Key activities related to energy infrastructure	**Conducts and Funds Technology Research** •Conducts research and development activities for the energy sector, including oil and gas, renewable ,and nuclear energy research •Technology development and deployment programs designed to modernize the U.S. electric delivery system **Protects Critical Infrastructure** •Applies DOE's technical expertise to ensure the security, resiliency and survivability of key energy assets and critical energy infrastructure at home and abroad **Collects and Analyzes Key Data on the Energy Sector** **Manages the Strategic Petroleum Reserve** **Administers the Power Marketing Administrations**[a]
Incorporation of climate change adaptation	•Supports and funds research to understand the impact of climate change on energy productionand to advance a predictive understanding of climate and inform development of sustainable solutions •Conducts assessments of climate change on electric grid stability, water availability for energy production, and site selection for the next generation of renewable energy infrastructure

Agency	**Department of Energy (DOE)**
Types of energy infrastructure impacted	Electricity transmission and distribution systems and oil storage
Key programs and offices	Office of Energy Policy and Systems, Office of Energy Efficiency and Renewable Energy, Office of Science, Office of Electricity Delivery and Energy Reliability, Strategic Petroleum Reserve, Energy Information Administration, and National Energy Technology Laboratory
Key legal authority for activities related to energy infrastructure	Energy Policy Act of 2005, Energy and Water Research Integration Act
Agency	**Environmental Protection Agency (EPA)**
Mission	Protects human health and the environment
Key activities related to energy infrastructure	**Regulates Hazardous Air Pollutants** •Emissions standards for power plants, petroleum refineries, and oil and gas extraction facilities **Regulates Waste Discharges into U.S. Waters** •Discharge and treatment of wastewater from power plants, petroleum refineries, and oil and gas extraction facilities **Responds to Oil Spills** **Regulates Cooling Water Intake Structures** •Power plant cooling systems **Regulates Solid and Hazardous Waste** •Fossil fuel combustion waste

Appendix II. (Continued).

Agency	**Environmental Protection Agency (EPA)**
	•Crude oil and natural gas waste **Prevents Contamination of Underground Drinking Water Resources** •Underground wells associated with natural gas and oil production **Oversees Greenhouse Gas Reporting Program** **Reviews Preconstruction Permits for Natural Gas Pipelines**
Incorporation of climate change adaptation	EPA has established agency-wide priority actions to begin integrating climate change adaptation into its programs, policies, rules and operations. The priorities are not specific to energy infrastructure, butsome priority actions may impact EPA's activities related to energy infrastructure as follows: •Integrating climate change trend and scenario information into EPA rulemaking processes •Factoring legal considerations into adaptation efforts—EPA may need to evaluate the legal basis for considering climate change impacts in setting standards or issuing permits under the Clean Air Act and Clean Water Act •Developing program and regional office implementation plans—each of the national program offices will develop its own plan that provides more detail on how it will integrate climate adaptation into its planning and work

Agency	**Environmental Protection Agency (EPA)**
Types of energy infrastructure impacted	Power plants, petroleum refineries, oil and gas extraction facilities, natural gas pipelines, and petroleum storage tanks
Key programs and offices	Office of Air and Radiation, Office of Policy, Office of Solid Waste and Emergency Response, Office of Water, and Underground Injection Control Program
Key legal authority for activities related to energy infrastructure	Clean Air Act, Clean Water Act, Resource Conservation and Recovery Act, Oil Pollution Act, and Safe Drinking Water Act
Agency	**Federal Energy Regulatory Commission (FERC)**
Mission	Assists consumers in obtaining reliable, efficient, and sustainable energy services at a reasonable cost through appropriate regulatory and market means
Key activities related to energy infrastructure	**Electricity** •Regulates wholesale sales of electricity and transmission of electricity in interstate commerce •Oversees energy markets and mandatory reliability standards for the bulk power system **Natural Gas** •Regulates interstate pipeline and storage facility siting, and abandonment •Regulates natural gas transportation in interstate commerce and establish rates, terms, and conditions for service **Liquefied Natural Gas (LNG)** •Oversees siting of new LNG terminals •Oversees proposals for and operation of LNG terminals

Appendix II. (Continued).

Agency	Environmental Protection Agency (EPA)
	Hydropower •Issues licenses for the construction of new hydropower projects and for the continuance of existing projects (relicensing) •Oversees ongoing project operations, including dam safety inspections and environmental monitoring **Oil** •Establishes reasonable rates for transporting petroleum and petroleum products by pipeline •Regulates oil pipeline companies engaged in interstate transportation •Establishes equal service conditions to provide shippers with equal access to interstate pipeline transportation
Incorporation of climate change adaptation	•Reviews LNG terminal projects to determine if terminal operators have accounted for hurricane and flooding impacts •Considers trends in historical hydrologic data as part of analysis of project operations and resource effects for hydropower facilities
Types of energy infrastructure impacted	Electricity transmission lines and facilities, interstate natural gas pipelines and storage facilities, LNGterminals, interstate oil pipelines, and hydropower plants
Key programs and offices	Office of Electric Reliability, Office of Energy Infrastructure Security, Office of Energy Policy and Innovation, Office of Energy Projects, and Office of Energy Market Regulation

Agency	**Environmental Protection Agency (EPA)**
Key legal authority for activities related to energy infrastructure	Federal Power Act of 1935, Natural Gas Act of 1938, Public Utility Holding Company Act of 2005, Energy Policy Act of 2005, Outer Continental Shelf Lands Act, and Interstate Commerce Act
Agency	**Nuclear Regulatory Commission**
Mission	Licenses and regulates the nation's civilian use of byproduct, source, and special nuclear materials toensure the adequate protection of public health and safety, promotes the common defense and security, and protects the environment
Key activities related to energy infrastructure	**Reactor Licensing** •Issues licenses for all commercially owned nuclear power plants that produce electricity in the United States. •After the initial license is granted, the license may be amended, renewed, transferred, or otherwise modified, depending on activities that affect the reactor during its operating life. **Power Uprates** •An amendment to an existing reactor operating license to allow a licensee to operate a reactor ata higher power level.
Incorporation of climate change adaptation	Although not in response to climate change: •Provides guidance for design requirements for hurricane wind speed protection at nuclear power plants •Requires the operating fleet of nuclear power plants to compare their existing designs with the new plant requirement for protecting against flooding hazards
Types of energy infrastructure impacted	Nuclear power plants

Appendix II. (Continued).

Agency	**Nuclear Regulatory Commission**
Key programs and offices	Office of Nuclear Reactor Regulation, Office of New Reactors, Regional Offices, Office of Nuclear Regulatory Research , and Advisory Committee on Reactor Safeguards
Key legal authority for activities related to energy infrastructure	Atomic Energy Act of 1954, Energy Reorganization Act of 1974
Nongovernmental organization	**North American Electric Reliability Corporation (NERC)**
Mission	Ensures the reliability of the North American bulk power system. NERC is the electric reliability organization certified by the Federal Energy Regulatory Commission to establish and enforce reliability standards for the bulk power system.
Key activities related to energy infrastructure	•Develops and enforces reliability standards •Assesses adequacy of generation and transmission annually via a 10-year forecast and winter and summer forecasts •Monitors the bulk power system and analyzes its performance •Analyzes system disturbances and distributes lessons learned •Operates Electricity Sector Information Sharing and Analysis Center •Educates, trains, and certifies industry personnel
Incorporation of climate change adaptation	•Reflects projections of weather and other variables that potentially impact bulk system reliability as part of its adequacy assessments. •Reports on system performance including demand response issues on a yearly basis. Weather forecasting uncertainties, in part reflecting climate change adaptation, are included as part of long-term load, generation, and transmission forecasting activities.

Nongovernmental organization	**North American Electric Reliability Corporation (NERC)**
Types of energy infrastructure impacted	Electric power generation facilities, high-voltage electric transmission lines and facilities, and transmission and generation control centers
Key programs and offices	Reliability Standards Development, Compliance and Enforcement, Critical Infrastructure Protection, Reliability Assessment and Performance Analysis, Reliability Risk Management, Event Analysis, andOperator Training
Key legal authority for activities related to energy infrastructure	Section 1211(a) of the Energy Policy Act of 2005 (Section 215 of the Federal Power Act)

Source: GAO analysis

[a]According DOE, its Power Marketing Administrations (PMA) provide electric power, largely hydropower from federal dams, to customers in 32 western, southwestern, and southeastern states and maintain an infrastructure that includes electrical substations, high-voltage transmission lines and towers, and power system control centers.

End Notes

[1] NRC is the operating arm of the National Academy of Sciences and National Academy of Engineering. Through its independent, expert reports; workshops; and other scientific activities, NRC's mission is to improve government decision making and public policy, increase public understanding, and promote the acquisition and dissemination of knowledge in matters involving science engineering, technology, and health.

[2] USGCRP coordinates and integrates the activities of 13 federal agencies that conduct research on changes in the global environment and their implications for society. USGCRP began as a presidential initiative in 1989, and the program was formally authorized by Congress in the Global Change Research Act of 1990 (Pub. L. No. 101-606, title I, 104 Stat. 3096-3104 (1990), codified at 15 U.S.C §§ 2931-2938). USGCRPparticipating agencies are the Departments of Agriculture, Commerce, Defense, Energy, Interior, Health and Human Services, State, and Transportation; the U.S. Agency for International Development; the Environmental Protection Agency; the National Aeronautics and Space Administration; the National Science Foundation; and the Smithsonian Institution.

[3] According to USGCRP, assessments assist decision making by surveying, integrating, and synthesizing science across sectors and regions. The key assessments used in this report compile information from many studies involving authors from academia; local, state, and federal government; the private sector; and the nonprofit sector. (See Objectives, Scope, and Methodology section for more information about the assessments used in this report.)

[4] Hurricane Sandy has also been popularly referred to as "Superstorm" Sandy due to the confluence of rare meteorological conditions that contributed to the widespread destruction of property and infrastructure in 2012.

[5] In response to extensive power outages during Sandy affecting millions of residents and resulting in substantial economic loss to communities, the federal government developed a Sandy Rebuilding Task Force that developed several recommendations regarding the alignment of investments in the Nation's energy infrastructure with the goal of improved resilience and national policy initiatives regarding climate change, transparency, and innovative technology deployment.

[6] Thomas R. Karl, Jerry M. Melillo, and Thomas C. Peterson, eds., *Global Climate Change Impacts in the United States* (New York, NY: Cambridge University Press, 2009), otherwise known as the 2009 National Climate Assessment.

[7] Since 1980, NOAA's NCDC has provided aggregated loss estimates for major weather and climate events, including tropical cyclones, floods, droughts, heat waves, severe local storms (tornado, hail, and wind damage), wildfires, crop freeze events and winter storms. The loss estimates reflect direct effects of weather and climate events and constitute total—insured and uninsured—losses. Specifically, estimates include physical damage to buildings; material assets; time element losses, such as hotel costs for loss of living quarters; vehicles; public and private infrastructure; and agricultural assets, such as buildings, machinery, and livestock. Estimates do not include losses to natural capital/assets, health care related losses, or values associated with loss of life. NOAA's NCDC defines climate as a statistical analysis of weather. Additional information available at NOAA's NCDC here.

[8] U.S. Climate Change Science Program (now known as USGCRP) *Draft Third National Climate Assessment Report,* Chapter 4 – Energy Supply and Use (January 2013). USGCRP, under the Global Change Research Act of 1990, periodically conducts a National Climate Assessment to inform the nation about observed climate changes and anticipated trends.

The Third National Climate Assessment is scheduled to be completed in early 2014; as a result, we have used the draft 2013 assessment for the purposes of this report. The draft 2013 assessment includes information from 240 authors drawn from academia; local, state, and Federal government; the private sector; and the nonprofit sector. The draft National Climate Assessment is not a finalized document and is subject to change as a result of comments from and review by the public, external entities, and the Federal Government. For more information or to access these assessments see http://www.globalchange.gov."

[9] See GAO, *Climate Change: Future Federal Adaptation Efforts Could Better Support Local Infrastructure Decision Makers*, GAO-13-242 (Washington, D.C.: Apr 12, 2013) and GAO, *Climate Change Adaptation: Strategic Federal Planning Could Help Government Officials Make More Informed Decisions*, GAO-10-113 (Washington, D.C.: Oct. 7, 2009).

[10] NRC, America's Climate Choices: *Panel on Adapting to the Impacts of Climate Change, Adapting to the Impacts of Climate Change* (Washington, D.C.: 2010).

[11] GAO, *High-Risk Series: An Update*, GAO-13-283 (Washington, D.C.: February 2013). Every 2 years at the start of a new Congress, GAO calls attention to agencies and program areas that are high risk due to their vulnerabilities to fraud, waste, abuse, and mismanagement, or are most in need of transformation.

[12] See GAO-13-283.

[13] According to USGCRP, assessments are tools to survey, integrate, and synthesize science. For more information about USGCRP assessments, click here. For objective 1 of this report, we generally used NRC's 2010 assessment, three USGCRP assessments (2007, 2009, and 2013), and DOE's assessments in 2010 and 2013, unless otherwise indicated. Citations for these assessments can be found in appendix I.

[14] We developed these four categories as a means of grouping similar processes together. Actual infrastructure and methods used to produce and distribute energy can vary.

[15] Because this was a nonprobability sample, findings from our examples cannot be generalized to all U.S. energy infrastructure; rather, they provide illustrative information about energy companies for which adaptation measures have been undertaken.

[16] When identifying agencies with key responsibilities related to energy infrastructure we focused on agencies with a direct role in overseeing and developing activities within the energy sector.

[17] NERC is not a federal agency; it is a nonprofit entity responsible for the reliability of the bulk power system (the generation and high-voltage transmission portions of the electricity grid) in North America (primarily the United States and Canada), but it is subject to the oversight of FERC and Canadian regulatory authorities. Because of NERC's important role with the electricity grid throughout the United States and oversight by FERC we will, hereafter, refer to NERC as a "federal entity."

[18] According to the Intergovernmental Panel on Climate Change, about 50 percent of carbon dioxide emitted by human activity will be removed from the atmosphere within 30 years, and a further 30 percent will be removed within a few centuries. The remaining 20 percent may stay in the atmosphere for many thousands of years. USGCRP estimates that another 0.5 degree Fahrenheit increase would occur even if all emissions from human activities were suddenly stopped.

[19] See GAO, *Technology Assessment: Climate Engineering: Technical Status, Future Directions, and Potential Responses*, GAO-11-71 (Washington, D.C.: July 28, 2011) for a depiction of the global carbon cycle changes over time and the global average "energy budget" of the Earth's atmosphere.

[20] President's Council of Economic Advisers, *Economic Benefits of Increasing Electric Grid Resilience to Weather Outages* (Washington, D.C.: August 2013).

[21] See GAO 13-242.

[22] Similarly, the Department of Homeland Security's recent 2013 National Infrastructure Protection Plan (Partnering for Critical Infrastructure Security and Resilience) defines resilience as the "ability to prepare for and adapt to changing conditions and withstand and recover rapidly from disruptions..."

[23] The IPCC is a scientific body under the auspices of the United Nations (UN). It reviews and assesses the most recent scientific, technical and socioeconomic information produced worldwide relevant to the understanding of climate change. It neither conducts any research nor monitors climate related data or parameters.

[24] According to a 2011 U.S. Geological Survey paper, these platforms were not designed to accommodate a permanent increase in sea level. See Burkett, Virginia, U.S. Geological Survey, "Global Climate Change Implications for Coastal and Offshore Oil and Gas Development," *Energy Policy* 39 (2011).

[25] U.S. Department of Energy, *Comparing the Impacts of Northeast Hurricanes on Energy Infrastructure* (April 2013).

[26] U.S. Department of Energy, *Comparing the Impacts of the 2005 and 2008 Hurricanes on U.S. Energy Infrastructure* (February 2009).

[27] By way of protection, the Alaska Department of Natural Resources limits the amount of travel on the tundra. Over the past 30 years, the number of days where travel is permitted has dropped from more than 200 to 100, thereby reducing by at least half the number of days that natural gas and oil exploration and extraction equipment can be used.

[28] Crude oil and petroleum products are transported by rail, barge systems, pipelines, and tanker trucks. Coal is transported by rail, barge, truck, and pipeline. Corn-based ethanol, blended with gasoline, is largely shipped by rail, while bioenergy feedstock transport relies on barge, rail, and truck freight.

[29] The Department of the Interior's bureaus are responsible for overseeing the processes that oil and gas companies must follow when leasing, drilling, and producing oil and gas from federal leases. The Minerals Management Service, a bureau within the Department of the Interior, was responsible for managing offshore activities and collecting royalties for oil and gas leases until May 2010, when the bureau was reorganized. Under this reorganization, the Bureau of Ocean Energy Management (BOEM) and the Bureau of Safety and Environmental Enforcement (BSEE) now oversee offshore oil and gas activities and the newly established Office of Natural Resources Revenue (ONRR) is responsible for collecting royalties on oil and gas produced from both onshore and offshore federal leases.

[30] U.S. Department of Transportation, *ExxonMobil Silvertip Pipeline Crude Oil Release into the Yellowstone River in Laurel, MT on 7/1/2011,* Pipelines and Hazardous Materials Safety Administration, Office of Pipeline Safety, Western Region.

[31] As permafrost thaws, the tundra loses its weight-bearing capabilities, according to DOE. Risks to onshore fossil fuel development could include the loss of access roads built on permafrost, loss of opportunities to establish new roads, problems with pipelines buried in permafrost, and reduced load-bearing capacity of buildings and structures.

[32] This improved accessibility will not be uniform throughout different regions, according to USGCRP, and extraction and exploration equipment may have to be redesigned to accommodate the new environment.

[33] Elcock, D., "Future U.S. Water Consumption: The Role of Energy Production," Journal of the American Water Resources Association, vol. 46, no. 3 (2010): 447-460.

[34] Water use by thermoelectric power plants can be generally characterized as consumption, withdrawal, and discharge. Water consumption refers to the portion of the water withdrawn that is no longer available to be returned to a water source, such as when it has evaporated. Water withdrawals refer to water removed from the ground or diverted from a surface water source—for example, an ocean, river, or lake—for use by the plant. For many thermoelectric power plants, much of the water they withdraw is later discharged, although often at higher temperatures. According to the U.S. Geological Survey (USGS), in terms of water withdrawal, thermoelectric power was the largest source of water withdrawals (49 percent) in 2005, followed by irrigation at 31 percent. The amount of water discharged from a thermoelectric power plant depends on a number of factors, including the type of cooling technology used, plant economics, and environmental regulations. Some "once-through" systems can harm aquatic life—such as fish, crustaceans, and marine mammals—by pulling them into cooling systems or trapping them against water intake screens. The habitats of aquatic life can also be adversely affected by warm water discharges.

[35] DOE, *U.S. Energy Sector Vulnerabilities to Climate Change and Extreme Weather*, DOE/PE-0013 (Washington, D.C.: July 2013).

[36] To prevent hot water from doing harm to fish and other wildlife, power plants typically are not allowed to discharge cooling water above a certain temperature. When power plants reach those limits, they can be forced to reduce power production or shut down.

[37] EPRI. 2011. *Water Use for Electricity Generation and Other Sectors: Recent Changes (1985–2005) and Future Projections* (2005–2030). 1023676. Palo Alto, CA: Electric Power Research Institute (November 10, 2011). http://www.epri.com/abstracts/Pages/Product Abstract.aspx?ProductId=000000000001023 676.

[38] See the "Water Availability" section of this section for further information on competing demands for water.

[39] According to NRC documents, Fort Calhoun remained closed as of November 1, 2013.

[40] Generally, under the Renewable Fuel Standard, which is overseen by EPA, transportation fuels in the United States are required to contain 36 billion gallons of biofuels annually by 2022.

[41] According to DOE, water use in biofuel refineries has been significantly reduced as a result of energy- and water-efficient designs in new plants and improved system integration in existing plants.

[42] According to DOE, CSP power plants using recirculating cooling water typically consume more water than a fossil fuel or nuclear power plants.

[43] Electricity generated through power plants or renewable energy sources is typically sent through high-voltage, high-capacity transmission lines to areas where it will be used; substations then transform the electricity to lower voltages and send it through local distribution wires to homes and businesses.

[44] Although wind-related outages do occur on transmission systems, about 90 percent of outages during a storm event occur along distribution systems, according to DOE.

[45] According to USGCRP, over the last century, snowstorms have increased in frequency in the Northeast and upper Midwest and decreased in frequency in the South and lower Midwest.

[46] Roberto Schaeffer, Alexandre Salem Szklo, André Frossard Pereira de Lucena, Bruno Soares Moreira Cesar Borba, Larissa Pinheiro Pupo Nogueira, Fernanda Pereira Fleming, Alberto Troccoli, Mike Harrison, Mohammed Sadeck Boulahya, "Energy Sector Vulnerability to Climate Change: A Review," *Energy* 38 (2012).

[47] DOE, Office of Electricity Delivery and Energy Reliability, *Comparing the Impacts of Northeast Hurricanes on Energy Infrastructure* (Washington, D.C.: April 2013).

[48] Since 2009, GAO has issued five reports on the interdependencies that exist between energy and water. *GAO, Energy-Water Nexus: Improvements to Federal Water Use Data Would Increase Understanding of Trends in Power Plant Water Use*, GAO-10-23 (Washington, D.C.: Oct. 16, 2009); GAO, *Energy-Water Nexus: Many Uncertainties Remain about National and Regional Effects of Increased Biofuel Production on Water Resources*, GAO-10-116 (Washington, D.C.: Nov. 30, 2009); GAO, *Energy-Water Nexus: Amount of Energy Needed to Supply, Use, and Treat Water Is Location-Specific and Can Be Reduced by Certain Technologies and Approaches*, GAO-11-225 (Washington, D.C.: Mar. 23, 2011); GAO, *Energy-Water Nexus: A Better and Coordinated Understanding of Water Resources Could Help Mitigate the Impacts of Potential Oil Shale Development*, GAO-11-35 (Washington, D.C.: Oct. 29, 2010); and GAO, *Energy-Water Nexus: Information on the Quantity, Quality, and Management of Water Produced during Oil and Gas Production*, GAO-12-156 (Washington, D.C.: Jan. 9, 2012).

[49] GAO, *Oil and Gas: Information on Shale Resources, Development, and Environmental and Public Health Risks,* GAO-12-732 (Washington, D.C.: Sept. 5, 2012). Water used in shale oil and gas development is largely considered to be consumptive and can be permanently removed from the hydrologic cycle, according to EPA and Interior officials. However, it is difficult to determine the long- term effect on water resources because the scale and location of future operations remains largely uncertain. Similarly, the total volume that operators will withdraw from surface water and aquifers for drilling and hydraulic fracturing is not known until operators submit applications to the appropriate regulatory agency. As a result, the cumulative amount of water consumed over the lifetime of the activity remains largely unknown.

[50] Water consumption is the portion of the water withdrawn that is no longer available to be returned to a water source, such as when it has evaporated. Energy production (which includes biofuel production), together with thermoelectric power, is the second largest consumer of water in the United States, accounting for approximately 11 percent of water consumption in 2005. Irrigation was the largest consumer, at approximately 74 percent. (Elcock, D., "Future U.S. Water Consumption: The Role of Energy Production," *Journal of the American Water Resources Association* vol. 46, no. 3 (2010): 447-460.). However, according to the U.S. Geological Survey, in terms of water withdrawal, thermoelectric power was the largest source of water withdrawals (49 percent) in 2005, followed by irrigation at 31 percent. Water withdrawal refers to water removed from the ground or diverted from a surface water source, such as an ocean, river, or lake.

[51] According to EPA, water from snowpack declined for most of the western states from 1950 to 2000, with losses at some sites exceeding 75 percent. Annual streamflows are expected to decrease in the summer for most regions, according to USGCRP, and drought conditions—which have become more common and widespread over the past 40 years in the Southwest, southern Great Plains, and Southeast, according to USGCRP— are expected to become more frequent and intense. Groundwater resources are already being depleted in multiple regions, according to USGS, and these impacts are expected to continue. See EPA, *Climate Change Indicators in the United States,* EPA 430-R-10-007 (Washington, D.C.: 2010) and United States Geological Survey, *Groundwater Depletion in the United States (1900–2008), Scientific Investigations Report 2013–5079* (Reston, VA: May 2013).

[52] Many factors can affect energy demand, including temperature and other weather conditions, population, economic conditions, energy prices, and conservation programs.

[53] I PCC, *Managing the Risks of Extreme Events and Disasters to Advance Climate Change Adaptation:* A Special Report of Working Groups I and II of the Intergovernmental Panel

on Climate Change [C.B. Field, V. Barros, T.F. Stocker, D. Qin, D.J. Dokken, K.L. Ebi, M.D. Mastrandrea, K.J. Mach, G.-K. Plattner, S.K. Allen, M. Tignor, and P.M. Midgley (eds.)]. (Cambridge, UK, and New York, NY : 2012).

[54] According to DOE, projected increases in air and water temperatures could significantly reduce electricity generation capacity, particularly in the summer months, by (a) decreasing the efficiency of power plant generation, (b) forcing power plant curtailments due to thermal discharge limits, (c) reducing electricity generated through hydropower and photovoltaic solar sources, and (d) increasing the temperature of local water sources to the extent they can no longer be used for cooling water.

[55] The energy supply chain is essentially a system of interconnected markets containing energy infrastructure that begins with extraction or generation of basic energy and ends with retail outlets for energy products. Along this chain, suppliers of inputs interact with demanders or consumers of these inputs, and both can avail themselves of substitute courses of action in adapting to climate change. For example, companies in the business of supplying oil to refineries may have alternative sources of oil, offshore and onshore that vary in their vulnerability to climate change impacts. In a mirror image, companies in the business of refining oil—consumers of oil as an input—may be able to avail themselves of these supply choices by switching their demand to less vulnerable oil.

[56] The Strategic Petroleum Reserve was authorized by Congress in 1975, following the Arab oil embargo of 1973-1974. It is owned by the federal government and operated by DOE. Under prescribed conditions, the President and the Secretary of Energy have the discretion to authorize release of oil in the Reserve through loans or other means to minimize significant supply disruptions and protect the economy from damage. [See GAO, *Strategic Petroleum Reserve: Available Oil Can Provide Significant Benefits, but Many Factors Should Influence Future Decisions about Fill, Use, and Expansion* GAO-06-872 (Washington, D.C.: Aug. 24, 2006).]

[57] DOE's Energy Information Administration, *New York/New Jersey Intra Harbor Petroleum Supplies Following Hurricane Sandy: Summary of Impacts through November 13, 2012* (Washington, D.C.: November 2012).

[58] In our illustrative example, choices in the gasoline supply chain represent investments that are already in place regardless of climate change. Therefore, what is relevant in considering the cost of using the substitutes described are the incremental costs of making greater use of these existing, or potentially new, substitutes compared with not adapting.

[59] AWF, established in Louisiana, and working throughout the Gulf region, was founded in 2002 in response to a comprehensive coastal study calling on the need to alert the nation to the devastating loss of Louisiana's coastal wetlands and how their loss impacts the rest of the nation. Coastal barriers and wetlands can help reduce infrastructure damage from weather events along the coast.

[60] In addition, Entergy has participated in Community Leadership Forums and technical conferences to educate local communities of potential climate change risks, help identify cost-effective measures to manage risk and discuss how the company could prioritize its investments in system hardening to minimize business interruption losses.

[61] The pipeline travels through the coastal states of Texas, Louisiana, Mississippi, Alabama, Georgia, South Carolina, North Carolina, Virginia, Maryland, Pennsylvania, and New Jersey. Branches from the main pipeline also reach Tennessee.

[62] Colonial representatives told us that the company sold the mobile generators used for power outages during Hurricanes Katrina and Rita in 2011. Since then, the company purchased eight new state-of-the-art mobile generators that were used in response to Hurricane Sandy.

[63] NRC requires the designs of structures, systems, and components important to safety to reflect appropriate consideration of the most severe natural hazards (NRC refers to natural hazards as natural phenomena) that had been historically reported for a reactor site and the surrounding area, with sufficient safety margin to account for the limited accuracy, quantity, and period over which historical data on natural hazards have been accumulated.

[64] Although most energy infrastructure is privately owned, the federal government owns a small number of hydroelectric power plants, transmission infrastructure, and strategic oil stock. The federal government's ownership of energy infrastructure is primarily managed through the Tennessee Valley Authority (TVA) and DOE through its Power Marketing Administrations (PMA) and Office of Petroleum Reserves. TVA supplies electricity to customers in parts of Tennessee, Alabama, Mississippi, Kentucky, Georgia, North Carolina, and Virginia. Similarly, three of the four DOE PMAs own and operate electricity transmission infrastructure. The PMAs distribute and sell electricity from a network of federally built hydroelectric dams, one nonfederal nuclear power plant, and several other small nonfederal power plants. DOE is also responsible for maintaining the infrastructure necessary to deliver crude oil from Strategic Petroleum Reserve.

[65] GAO-13-283.

[66] The El Niño Southern Oscillation is a natural occurring phenomenon that involves fluctuating ocean temperatures in the equatorial Pacific Ocean. The pattern generally fluctuates between two states: warmer than normal temperatures in the central and eastern equatorial Pacific (El Niño) and cooler than normal temperatures in the central and eastern equatorial Pacific (La Niña).

[67] USGCRP, *The National Global Change Research Plan 2012-2021, A Strategic Plan for the U.S. Global Change Research Program* (2012).

[68] USGCRP is developing the Global Change Information System (GCIS), a comprehensive web-based system to deploy and manage global change information. This system will support the NCA by producing reports while also incorporating integrated and linked access to data to ensure open and transparent access to climate information.

[69] GAO, *EPA Regulations and Electricity: Better Monitoring by Agencies Could Strengthen Efforts to Address Potential Challenges*, GAO-12-635 (Washington, D.C.: July 17, 2012).

[70] NERC cannot require energy infrastructure owners to make a specific technological change, enlarge their facilities or construct new transmission or generation capacity, according to NERC officials. The officials stated that the specific actions that industry takes to meet NERC's standards can vary.

[71] According to DOE, Smart Grid technology means "computerizing" the electric utility grid. It includes adding two-way digital communication technology to devices associated with the grid. Each device on the network can be given sensors to gather data (power meters, voltage sensors, fault detectors, etc.), plus two-way digital communication between the device in the field and the utility's network operations center. A key feature of the smart grid is automation technology that lets the utility adjust and control each individual device or millions of devices from a central location.

[72] NFIP is administered by FEMA. The Congressional Budget Office (CBO), the Council of Economic Advisers (CEA), and GAO have cited insurance premiums as not fully reflecting risks. (See: CBO, "The National Flood Insurance Program: Factors Affecting Actuarial Soundness, " November 2009; CEA, "Economic Report of the President," March 2013; and GAO, *Climate Change: Financial Risks to Federal and Private Insurers in Coming Decades Are Potentially Significant,* GAO-07-285 (Washington, D.C.: Mar. 16, 2007); and GAO, *FEMA: Action Needed to Improve Administration of the National Flood Insurance*

Program, GAO-11-297 Washington, D.C.: June 9, 2011). To address this issue, the Biggert-Waters Flood Insurance Reform Act of 2012 (Pub. L. No. 112-141, title II, Jul. 6, 126 Stat. 916 (2012), (codified at 42 U.S.C. § 4001–4129)) proposes steps to address shortcomings in the NFIP by authorizing FEMA to consider information such as changing coastal topography, erosion rates, sea level rise projections, and changes in intensity of hurricanes in its future flood maps. These changes are expected to significantly increase NFIP's insurance premium rates, in some cases.

[73] Lower quality water generally can refer to degraded or nonpotable water such as contaminated groundwater, treated municipal wastewater, industrial process water, irrigation return water, brackish water, and other types of water impacted by humans or naturally occurring minerals.

[74] In commenting on this report, the Nuclear Regulatory Commission clarified that it does not directly regulate the impact of climate change on nuclear power plants but requires these plants to be protected against the effects of certain natural phenomena. Thus, in as much as climate change affects these natural phenomena, the Commission believes climate change impacts are taken into consideration in its review process.

[75] On November 1, 2013, President Obama signed Executive Order (EO) 13653 on preparing the United States for impacts of climate change. The EO directs U.S. federal agencies to take steps that will make it easier for American communities to strengthen their resilience to extreme weather and to prepare for other impacts of climate change. As a result, the Administration established the Council on Climate Preparedness and Resilience, an interagency working group. The Council includes an Infrastructure Working Group, co-chaired by the Department of Energy and the Department of Homeland Security that focuses on infrastructure resilience to climate change.

[76] FERC and NRC have not developed agency adaptation plans. FERC officials told us that FERC is not subject to the requirement to develop an agency climate change adaptation plan under Executive Order 13514. NRC officials told us that the Council on Environmental Quality (CEQ) approved of NRC using its climate change adaptation policy statement in lieu of developing an adaptation plan. NERC is not a federal agency and, therefore, was not required to develop a climate change adaptation plan.

[77] The task force, which began meeting in Spring 2009, is co-chaired by CEQ, NOAA, and the Office of Science and Technology Policy and includes representatives from more than 20 federal agencies and executive branch offices, including DOE and EPA. The task force was formed to develop federal recommendations for adapting to climate change impacts both domestically and internationally and to recommend key components to include in a national strategy. FERC, NRC, and NERC are not members of the task force.

End Note for Appendix I

[1] NERC is a not-for-profit corporation that has been certified by FERC as the Electric Reliability Organization in accordance with section 1211(a) of the Energy Policy Act of 2005. As such, NERC is not a federal agency (see section 1211(b)), but we have included it in our discussion of federal action because it is responsible for developing and enforcing reliability standards for the North American bulk power system (the generation and high-voltage transmission portions of the electricity grid) and is subject to oversight by FERC within the United States and by Canadian regulatory authorities.

In: U.S. Energy Infrastructure
Editor: Joanne R. Ballard

ISBN: 978-1-63482-286-2
© 2015 Nova Science Publishers, Inc.

Chapter 2

DOE CLIMATE CHANGE ADAPTATION PLAN*

U.S. Department of Energy

CLIMATE CHANGE ADAPTATION PLAN

> "The threat from climate change is real and urgent. The science fully demands a prudent response." - Secretary of Energy, Ernest Moniz

Changes in the global climate system are unmistakable, as is now evident from observations of increased global average air and ocean temperatures, decreased historical snow pack, rising global average sea level, and more frequent severe weather events.[1] The U.S. Department of Energy (DOE) recognizes that changes in the global climate system could have a profound impact on the Department's mission activities. DOE is committed to reducing greenhouse gas (GHG) emissions and mitigating climate change by developing clean energy and energy efficiency technologies for commercial deployment while providing leadership through its own sustainable operations. As effects of climate change are felt across the world, it is necessary to characterize potential impacts on the DOE mission, programs, and operations to foster adaptation and resilience. DOE will identify where to focus resources to develop greater resilience over time, minimize potential risks and maximize potential opportunities created by climate change. It is important to note that

* This is an edited, reformatted and augmented version of a document issued June 2014.

while longer term damages could be very substantial, we may have more modest, nearer term impacts (above and beyond those that we already have due to weather vulnerabilities). The 2014 DOE Climate Change Adaptation Plan (Adaptation Plan) outlines Departmental vulnerabilities and serves to guide our response to allow DOE to continue to achieve its mission.

The DOE vision for climate change adaptation is the integration of risk based resiliency to address identified climate change vulnerabilities across all DOE programs and policies wherever appropriate. Assessment of climate change vulnerabilities, informed by best available science, are seen as an integral part of the DOE's planning activities, risk assessment, and careful investment that define the DOE mission.

Climate change adaptation is not new to DOE; rather, climate change is an ongoing part of DOE research, modeling, and policy development. A strong culture of preparedness, integrated safety management, and operational excellence in potentially hazardous working environments already exists throughout DOE. Climate change resilience will build on this operational and capital planning, as well as provide information to the larger applied research body of climate change adaptation.

Impetus for Action

The Adaptation Plan serves as the second iteration of a living plan. This edition was modified in accordance with Executive Order (E.O.) 13653, *Preparing the United States for Climate Change*, the knowledge gained from our first Adaptation Plan (publicly released in 2012), and the experiences of the Department and other federal agencies in response to extreme events such as Hurricane Sandy. The Department's Adaptation Plan will also draw from E.O. 13514, *Federal Leadership in Environmental, Energy, and Economic Performance*, and the President's Climate Action Plan (June 2013) to provide a comprehensive planning document.

The Adaptation Plan addresses the national and international context of the DOE mission, as well as the local perspective of DOE facilities and community stakeholders. Increased understanding of climate change has enabled DOE to better forecast climate change impacts, quantify risk, and identify opportunities to improve resilience.

The Adaptation Plan will work in concert with DOE's ongoing mitigation activities outlined in DOE's annual Strategic Sustainability Performance Plan

(SSPP)[2]. DOE maintains its commitment to reducing agency GHG emissions, utilizing renewable energy, and making operations more sustainable.

The DOE Mission and Climate Change Adaptation

DOE's mission is to "enhance U.S. security and economic growth through transformative science, technology innovation, and market solutions to meet our energy, nuclear security, and environmental challenges."[3] DOE achieves its mission through an operational and programmatic framework that supports the following goals outlined in the 2014-2018 DOE Strategic Plan:

- Advance foundational science, innovate energy technologies, and inform data driven policies that enhance U.S. economic growth and job creation, energy security, and environmental quality, with emphasis on implementation of the President's Climate Action Plan to mitigate the risks of and enhance resilience against climate change
- Strengthen national security by maintaining and modernizing the nuclear stockpile and nuclear security infrastructure, reducing global nuclear threats, providing for nuclear propulsion, improving physical and cybersecurity, and strengthening key science, technology, and engineering capabilities
- Position the Department of Energy to meet the challenges of the 21st century and the nation's Manhattan Project and Cold War legacy responsibilities by employing effective management and refining operational and support capabilities to pursue departmental missions.

The DOE enterprise is comprised of approximately 15,000 federal employees and over 100,000 contractor employees at both the headquarters in Washington, DC and Germantown, Maryland and at over 47 facilities in 40 states (see Figure 1). DOE's facilities are located in all eight U.S. climate regions identified in the 2014 National Climate Assessment, as established by the U.S. Global Change Research Program (USGCRP).[4] These facilities include a nationwide system of 17 national laboratories that provide world-class scientific, technological, and engineering capabilities that employ over 29,000 researchers. These laboratories include operate numerous scientific user facilities that allow researchers from other Federal agencies, universities and the private sector to use. DOE facilities also contain unique scientific equipment and processes and house advanced materials, including nuclear materials critical to U.S. national security.

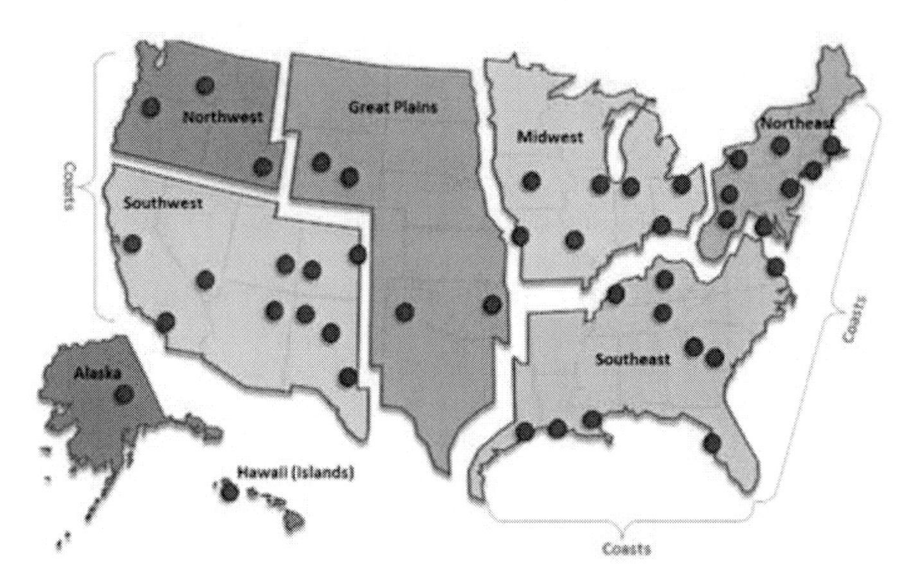

Figure 1. DOE Sites Grouped in USGCRP Climate Regions.

DOE leads the nation in transformational research, development, demonstration, and deployment (RDD&D) of an extensive range of clean energy and efficiency technologies – supporting the President's Climate Action Plan and an all-of-the-above energy strategy. DOE also leads national efforts to develop technologies to modernize the electricity grid, enhance the security and resilience of energy infrastructure, and expedite recovery from energy supply disruptions. The Department conducts robust, integrated policy analysis and engagement in support of the nation's energy agenda.

DOE enhances the security and safety of the nation through its national security endeavors: maintaining a safe, secure, and effective nuclear weapons stockpile in the absence of nuclear testing and managing the research, development, and production activities and associated infrastructure needed to meet national nuclear security requirements; accelerating and expanding efforts to reduce the global threat posed by nuclear weapons, nuclear proliferation and unsecured or excess nuclear materials; and, providing safe and effective nuclear propulsion for the U.S. Navy.

DOE leads one of the largest cleanup efforts in the world to remediate the environmental legacy of over six decades of nuclear weapons research, development, and production. As DOE carries out its mission, it employs effective and cost-efficient management, supports a highly skilled workforce, and provides a modern physical and information technology infrastructure. DOE remains committed to maintaining a safe and secure work environment for all personnel and to ensuring that its operations preserve the health, safety, and security of the surrounding communities.

The distribution of DOE facilities across all eight U.S climate regions gives DOE a unique opportunity to be a leader in climate change assessments and resiliency efforts. DOE's 2014-2018 Strategic Plan highlights the Department's leadership and commitment to bringing the extraordinary technical resources of the Department to bear in this effort. The DOE vision for climate change adaptation is the integration of climate change resiliency across the Department wherever appropriate. DOE will look to incorporate the risks and impacts posed by climate change in DOE plans, programs and policies, and mission.

Risks to Operations, Missions and People

Findings of the DOE High-Level Vulnerability Assessment

DOE completed a high level vulnerability assessment in 2012 to determine departmental risks to climate change[5]. Locations where the DOE mission is conducted were found to be likely to encounter a combination of climate change effects, both in the form of direct acute events (e.g., severe weather) and long-term changes (e.g., average annual precipitation). Additional indirect impacts, particularly those related to social systems and human health, were also recognized. These were found to possibly manifest as both acute (e.g., economic and political instability) and long-term (e.g., changes in population and social demographics) vulnerabilities. DOE identified the following high level critical vulnerabilities:

- DOE could potentially be exposed to global, national, regional and location-specific effects of climate change;
- DOE's mission, site operations and programs have varying degrees of sensitivity, depending on location and type of work. This diversity must be accounted for in the planning process in order to prevent maladaptation.
- Given the nature of its mission and in-house expertise, DOE possesses considerable adaptive capacity within existing policy, planning, and

operational frameworks. Greater and more consistent integration can develop resilience to future climate change effects.

DOE has the opportunity to enhance its own operations and programs through increased climate change resiliency, while also contributing technical expertise and climate change resilient energy technology solutions. The Department can therefore both provide continuity for its own operations and help the nation and its international partners to adapt to a changing climate.

Physical Assets and Real Property

DOE physical assets and real property are diverse and vary in complexity. The Department's assets include typical office buildings, laboratory buildings, warehouses and shops, as well as unique one-of-a-kind research facilities such as light sources, research reactors, semi-conductor fabrication facilities, particle accelerators, and nuclear material production and processing facilities. This diversity of facilities results in a multitude of vulnerabilities, risks and impacts.

The Energy-Water Nexus

The energy-water nexus poses a significant risk to DOE sites and operations. Extreme climate events such as increased temperature and associated drought have important impacts on the energy-water interface. Climate impacts will manifest as changes in land use and resource demand. These changes tend to reinforce and intensify individual impacts on land and water resources (e.g., reduced cropping due to water shortages raises feed prices, which changes grazing patterns, which in turn affects vegetation density and thus potentially increases wild fire vulnerability). To an extent, these changes feedback through water and land use to impact energy demand and production[6].

In FY 2013, DOE facilities used 4.947 million megawatt hours (MWh) of electricity. Much of this electricity is procured by DOE from off-site providers that utilize hydropower or steam electric power generation requiring significant volumes of water. Interruption of electricity for any extended period of time will greatly impact DOE mission activities. Examples of how this could occur are discussed below.

Warmer temperatures and changes in the hydrologic cycle, including precipitation type, frequency, and intensity are expected with climate change –

putting generating capacity at risk. Temporally, this impact may cause concern with short term, seasonal fluctuations as well as longer term disruptions in service capacity. The Power Marketing Administrations within DOE operate federal hydropower facilities and may be subject to additional vulnerabilities as well. The potential for changes in reservoir operations to meet new, non-power uses, and the aging of federal hydropower assets, could lead to lower reliability and a reduction in operating capacity[7].

Climate change impacts to water availability may impact energy availability at other DOE facilities as well. Water accessibility for commercial offsite electricity production may impact electricity production and energy costs. The increased incidence of droughts and water constraints could result in water-related electricity shortages and disruption of DOE facility operations.

Water Vulnerabilities

Declines in the quantity and quality of water associated with climate change may directly affect site operations and research efforts. In FY 2013, DOE used 6.464 billion gallons of potable water, and an additional 1.171 billion gallons of non-potable water. Facility and industrial uses account for the majority of DOE's potable and non-potable water use. The availability of significant volumes of water will remain essential to critical facility operations.

Prolonged droughts coupled with warmer weather may also contribute to increased wildfires. Particularly in western regions of the country wild fire risk can pose a serious threat to facilities and local infrastructure supporting DOE operations. Wildfire damage has already proven a risk to DOE facilities (see Case Study 1) and may increase with climate change.

The amount of rain falling in extreme precipitation events has increased in the U.S. by approximately 20 percent in the past century, and this trend is projected to continue. A national upward trend in the number of extreme precipitation events is statistically significant with the greatest frequencies occurring in recent years[8]. More intense precipitation or storm events could result in flooding. DOE sites already experience floods under existing climate conditions (see Case Study 2) along with record breaking rainfalls, resulting in millions of dollars in damages.

Temperature Extremes

Increased average temperatures will result in decreased heating needs during winter months; they will also result in increased summer cooling needs for DOE buildings.

DOE Case Study 1: 2011 Wildfire at Los Alamos National Laboratory

Wildfires already pose a risk to DOE facilities in the Southwest. On June 26th, 2011, a tree falling on a power line ignited the Las Conchas wildfire in the Jemez Mountains near Los Alamos National Laboratory (LANL). This fire burned 156,593 acres before it was contained—the largest wildfire at the time in New Mexico's recorded history.

Improvements made based on lessons learned from the 2001 Cerro Grande Fire, such as improving fire protection systems across the Laboratory, undertaking massive fuel thinning projects, and improving fire breaks and roads, prepared LANL for the Las Conchas wildfire. LANL's high level of preparedness helped to protect a critical DOE asset from wildfires. As the effects of climate change continue to impact our sites LANL and DOE will have to adapt to changing ecosystems and fire regimes to ensure the asset remains protected

DOE Case Study 2: 2010 Extreme Flooding at Pantex

DOE's Pantex Plant is the nation's only nuclear weapons assembly and disassembly facility. Work performed at Pantex is critical to national security. The plant is located in flat terrain in the Texas Panhandle, with shallow and gentle slopes for drainage.

In July, 2010, the region experienced record breaking rainfall and severe flash flooding. Pantex received 11.04 inches of rain in a single day, 8.76 inches of which fell during a single hour. Natural and man-made drainage systems on the site were overwhelmed. Since this incident, Pantex has constructed improved drainage ditches on the site, and also response plans, procedures and equipment to better prepare for flash flooding events. As this example shows DOE is already experiencing increased frequency and intensity of climate related impacts. This adds increased urgency to adaptation efforts.

In addition, DOE sites could face operationally disruptive electricity shortages during peak summer demand periods due to increased DOE and non-DOE electricity demand. Other regions of the country may face more extreme climate events (extreme heat and cold) and more variable temperatures throughout the year. Results could include increased energy and heating/cooling costs and the need for additional capacity. Increased incidence of heat and climate related workforce health issues related to heat stress, potential increases in pestilent species or epidemics, and outdoor human operational constraints may affect day to day operations.

Sea Level Rise and Extreme Events

Climate change projections include sea level rise and a greater likelihood of severe weather events (e.g., hurricanes, storm surge, etc.). The effects of sea-level rise on coasts will vary considerably from region-to-region and over a range of spatial and temporal scales. Land subsidence in certain locations causes relative sea-level rise to exceed global mean sea-level rise. The effects will be greatest and most immediate on low-relief, low-elevation parts of the U.S. coast along the Gulf of Mexico, Mid-Atlantic States, northern Alaska, Hawaii, and island territories and especially on coasts containing deltas, coastal plains, tidal wetlands, bays, and estuaries.[9]

Even though a majority of DOE facilities are located inland, sea level rise and extreme events could impact those facilities located in coastal regions through inundation, especially during storm events, and salt water intrusion affecting freshwater availability and quality. In the longer term, coastal erosion could also increase vulnerability at some locations. Additionally, in a more regional context, inland DOE sites are staffed and supported by resources located in coastal locations.

Sea level rise may also affect external facilities that provide DOE with electricity. Most saltwater consumption in U.S. coastal counties occurs during thermoelectric power generation. Changes in water temperature, density and level may impact the effectiveness of water as a cooling medium. Additionally the coasts are areas of exploration for energy sources including traditional sources, such as the extraction and transportation of offshore fossil fuels to inland areas, and alternative sources, such as tidal, wave, and wind energy. DOE facilities that rely on external power sources will need to account for these vulnerabilities.

Appendix 1, *Summary of Forecasted Climate Effects for Major US Climate Regions*, illustrates examples of the effects of climate change in each major climate region. This table includes general regional effects summarized by the USGCRP with a focus on those most likely to impact DOE. Identifying the likelihood, timing, frequency, and magnitude of climate change impacts including secondary risks for specific DOE sites and their surrounding communities requires further analysis, outreach, research and modeling.

Programmatic Risks

Given that much of DOE's work is tied to critical specialized infrastructure, it is impossible to completely separate DOE's programs from location-specific facility operations. However, it is important to consider climate change not just in terms of infrastructure integrity and facility operations, but also in terms of broader programs and mission. DOE is working in its current programmatic structure to account for vulnerabilities and impacts are at more than just the for the majority of DOE's potable and non-potable water use. The availability of significant volumes of water will remain essential to critical facility operations.

Prolonged droughts coupled with warmer weather may also contribute to increased wildfires. Particularly in western regions of the country wild fire risk can pose a serious threat to facilities and local infrastructure supporting DOE operations. Wildfire damage has already proven a risk to DOE facilities (see Case Study 1) and may increase with climate change.

The amount of rain falling in extreme precipitation events has increased in the U.S. by approximately 20 percent in the past century, and this trend is projected to continue. A national upward trend in the number of extreme precipitation events is statistically significant with the greatest frequencies occurring in recent years[8]. More intense precipitation or storm events could result in flooding. DOE sites already experience floods under existing climate conditions (see Case Study 2) along with record breaking rainfalls, resulting in millions of dollars in damages.

Temperature Extremes

Increased average temperatures will result in decreased heating needs during winter months; they will also result in increased summer cooling needs for DOE buildings. In addition, DOE sites could face operationally disruptive electricity shortages during peak summer demand periods due to increased DOE and non-DOE electricity demand. Other regions of the country may face more extreme climate events (extreme heat and cold) and more variable temperatures throughout the year. Results could include increased energy and heating/cooling costs and the need for additional capacity. Increased incidence of heat and climate related workforce health issues Climate change may also affect DOE's ongoing role in preserving historic and cultural resources on its federal lands, partnerships with Tribal governments, and environmental justice responsibilities. DOE-managed historical and cultural resources will be exposed to the same climate change effects as adjacent DOE facilities which could present new challenges in protecting, preserving, curating, and mitigating impacts to these resources. In particular, the long-term integrity of DOE landfills and waste-storage sites containing hazardous materials could be affected by climate change, resulting in secondary impacts in adjacent or downstream culturally or environmentally sensitive areas, including Tribal lands. DOE is committed to implementing its Environmental Justice Strategy which reflects a commitment to the fair treatment and meaningful involvement of all people in agency programs. Moreover, the Memorandum of Understanding concerning E.O. 12898, *Federal Actions to Address Environmental Justice in Minority Populations and Low-Income Populations,* to which DOE is a party, specifically cites climate change as an area of focus when considering environmental justice[10].

Human Health and Safety

DOE sites are likely to encounter new climate change-related workforce challenges in the future, regardless of region. DOE sites may experience increased incidence of worker illness due to changing environmental conditions. Shifts in climate can change disease vectors that increase infectious disease exposures and may exacerbate existing health problems[11]. Other facilities may experience human welfare and quality of life issues caused by frequent or prolonged heat waves and heat stress. Heat related issues may result in vulnerabilities for employees both on and off the DOE complex. Potential local climate change impacts on workforce health and quality of life may affect the operational capacity of DOE's facilities. In all instances DOE sites will have to adapt operations to accommodate changes in workforce needs and safety concerns.

Another key aspect to consider is transportation infrastructure connecting DOE facilities with each other, the supply chain, and the DOE workforce. Roadways and mass transit infrastructure may be damaged or obstructed by extreme weather events, disrupting operations and endangering employees. Additionally, stress on public transportation infrastructure by climate change may impact operations. DOE will need to coordinate with neighboring communities to address effective adaptation and mitigation responses.

Building Resilience Current Activities

DOE has an array of programs, policies and plans currently in place that are beginning to address climate vulnerabilities across the enterprise. These activities detail a strong commitment by all facets of DOE, ensuring the Department and the nation is more resilient to climate change effects.

Site Sustainability Planning includes Climate Change Resilience Goal

DOE sites engage the Department's sustainability process and requirements through annual site specific reporting. Site Sustainability Plans (SSPs), driven primarily by E.O. 13514 and the Department's SSPP[12], have been leveraged to include adaptation activities. These efforts address facility, programmatic and human health vulnerabilities. DOE, through the Sustainability Performance Office, tracks site performance with respect to the goals outlined in the 2012 Adaptation Plan as well as the strategies outlined in the climate change resilience goal in the SSPP. These actions will continue to be measured on an annual basis through this process. Some of the strategies and corresponding activities include:

- Review DOE Orders, Guides and Technical Standards to determine which require updates, and prioritize updates on a multi-year schedule.
  - DOE identified DOE Guide 413.3-6A, *High Performance Sustainable Buildings,* as a planning guide that requires an update to include climate change resiliency language. The guide provides implementing instructions for implementing the HPSB requirements of DOE Order 413.3B, *Program and Project Management for the Acquisition of Capital Assets.* The update will also contribute to the agency's ongoing effort to consider climate adaptation and resilience into procurement, real property and leasing decisions. Additional opportunities to further this priority are under consideration.
  - DOE also identified DOE Order 150.1A, *Continuity Programs,* as an ideal opportunity to integrate climate change adaptation. This order outlines the requirements for DOE sites when completing continuity planning. Working with the Office of Emergency Management, DOE was able to include climate change in the list of potential risks and as a part of the All Hazard Risk assessment process. Finally E.O. 13653 was added as a reference for this

process. This update and future similar orders will help DOE sites to retain operational abilities in the face of emergencies associated with climate change.

- Identify DOE program and site specific planning documents and guidance that should be developed or updated to improve climate change resiliency. For example:
  - The Pacific Northwest National Laboratory's (PNNL) Sustainability Program identified and reviewed site plans that either currently address or present an opportunity to address climate change adaption including the Building Emergency Plan, Business Continuity Plan, and Campus Master Plan. The Campus Master Plan was determined to present the most important opportunity to address climate adaptation planning. The current plan does not specifically address climate change adaptation but does include commitments to climate change mitigation through sustainable campus design. The Plan is scheduled for revision in FY 2014, and the campus planning team is now committed to working with the Sustainability Program to understand and address the greatest vulnerabilities. PNNL will look to focus particularly on facility energy shortages and reduced water supply. These efforts address multiple departmental vulnerabilities including facility and programmatic risks.
- Strengthen and broaden the internal DOE Adaptation Working Group to involve sites, share experiences (e.g., Hurricane Sandy – see below), best practices, case studies and innovation across programs and sites, identify corporate and other federal resources to support site vulnerability assessment, and provide communication tools (e.g., web sites, video and teleconferences, newsletters such as the DOE Sustainability SPOtlight).
  - In October 2012, Brookhaven National Laboratory (BNL) in Upton, New York experienced the effects of a record storm event, Hurricane Sandy, which impacted most of the northeast. Significant upfront planning was initiated for this storm to understand building vulnerabilities and potential effects on the science mission. Directly after the storm, Preliminary Damage Assessment (PDA) Teams were directed by the Office of Emergency Management to assess the site and restore services. As a direct result of planning and coordination, the overall impact to BNL was minimal and ongoing projects were not significantly

impacted. After the storm, BNL re-opened for business earlier than planned, which provided a significant cost savings to BNL and DOE, and enabled researchers to continue their experiments. BNL"s planning and response serve as best practice for DOE sites preparing for extreme weather events.

- Continue group participation at DOE and inter-agency level, to ensure awareness of current and best available climate science and associated technologies with dissemination via the DOE Adaptation Working Group.

Activities to meet these objectives are occurring throughout the Department. Sustainability planning and reporting will continue to be leveraged to encourage further action in support of this plan. The sharing of best practices and success stories will enable DOE to meet the vision of agency wide resilience to climate change.

Regional Collaboration and Interagency Cooperation

The 2012 Adaptation Plan prioritized site-level regional coordination efforts and interagency cooperation to advance climate preparedness at DOE facilities. DOE looks to incorporate local and regional collaboration at all levels of climate change planning in recognition of the fact that vulnerabilities impacting our facilities also threaten their neighboring communities. Likewise, DOE will learn from local and regional efforts. DOE has incorporated regional collaboration into annual sustainability reporting collected by the Sustainability Performance Office. Examples of collaboration include:

- Oak Ridge National Laboratory, in collaboration with Sustainable Tennessee, produced the report "The State of the Future for a Sustainable Tennessee: Grand Challenges and Grand Opportunities Under Changing Climate" in August 2012[13]. The goal of the report was to outline vulnerabilities and pathways for the state moving forward.
- PNNL operated the Joint Global Change Research Institute (JGCRI) along with the University of Maryland. The institute houses an interdisciplinary team dedicated to understanding the problems of global climate change and their potential solutions. Research areas include carbon cycle science, climate change assessment modeling, energy technology, and adaptation and mitigation programs.

- Idaho National Laboratory (INL) serves as coordinating partner for the Mountain West Water Institute (MWWI) as part of the US Climate Change Technology Program, a multi-agency research program. INL has worked closely with NASA, the Pacific Northwest Collaboratory and the North Olympic Peninsula to develop decision support tools. The goal of MWWI's climate change work is to collaborate with state, regional and federal agencies, universities, and other researchers and stakeholders to develop a better understanding of the probabilities, vulnerabilities and potential impacts of projected climate changes. It will also develop strategies to avoid, adapt to or mitigate negative impacts or to take advantage of positive impacts relative to water resources. Furthermore, it is MWWI's goal to conduct such evaluations in a systematic and interdisciplinary manner, and to develop a holistic understanding and comprehensive response to potential vulnerabilities and impacts on the region's water resources.
- Thomas Jefferson National Accelerator Facility is a partner in the development of a Hampton Roads Energy Corridor. Hampton Roads is unique in its vast array of distributed Federal facilities and on site expertise in various forms of alternative energy and renewable resources. The Energy Corridor serves to draw on this expertise to create a more resilient region. Climate change vulnerabilities and adaptation are key planning components of this effort.

These examples and others are shared and discussed via the DOE Adaptation Working Group.

In addition to initiatives occurring at the Department's National Laboratories and facilities, DOE is also building resilience in communities impacted by Hurricane Sandy. The Office of Electricity and Energy Reliability is supporting the Sandia National Laboratories (SNL) to aid the city of Hoboken, NJ in boosting the resiliency of its electric grid[14]. This critical partnership brings the deep expertise of the national laboratories to address the critical needs of our nation's electric grid. SNL will bring its Energy Surety Design Methodology to partner with the City of Hoboken, New Jersey Board of Public Utilities, Public Service Electric and Gas Company (PSE&G), Greener by Design and other stakeholders to develop a comprehensive plan to meet the critical needs of Hoboken in future events such as storms and other disruptions to the electric grid. The design methodology uses advanced, smart grid technologies and distributed and renewable generation and storage resources as a way to improve the reliability, security, and resiliency of the electric grid. Activities such as this allow DOE to engage local communities on improving the resilience of the energy sector in support of the DOE mission.

Site-Level Climate Vulnerability Assessments

In 2013, DOE began work on a series of pilot site-level climate vulnerability assessments to develop tools and templates that would be useful to other DOE sites. These assessments engage sites on climate change risk at a level of detail useful for site planning and decision making. The projects are designed to address a broad range of climate-related risks at DOE facilities. This work is intended to be a platform for further assessment for other sites and will attempt to standardize climate change risk assessment approaches and processes across missions and operations. The results will be used to engage decision makers and support preparation of future site planning documents. Initial stages of the project will be completed by the end of 2014.

The first pilot site chosen was the Thomas Jefferson National Accelerator Facility (TJNAF) due to its coastal location in Newport News, Virginia. This area is particularly vulnerable to coastal storms and sea level rise. The assessment and corresponding work will provide DOE with an opportunity to not only engage our own facilities but also the larger federal community collocated in the Hampton Roads area. DOE will look to partner with other agencies and local communities, where possible, to develop stronger planning and increase resiliency.

The second pilot site chosen was the National Renewable Energy Laboratory (NREL) in Golden, Colorado. NREL is situated in an inland climate and will provide DOE an additional perspective on the vulnerability assessment process. NREL also has close ties to its local community and surrounding federal facilities and will look to develop partnerships to increase adaptive capacity and resilience.

Energy Sector Vulnerabilities to Climate Change and Extreme Weather

In 2014, DOE released an assessment of climate change impacts on the United States energy sector. The assessment included a review of key vulnerabilities – including power plant disruptions due to drought and disruption of fuel supplies during severe storms – as well as a discussion of potential opportunities to make our energy infrastructure more resilient to climate risks. The DOE report, *US Energy Sector Vulnerabilities to Climate*

Change and Extreme Weather, covers all aspects of the energy system, including supply, distribution and end use, and provides a comprehensive, high-level examination of impacts related to increasing temperatures, decreasing water availability, and increasing intensity of storms, flooding, and sea level rise.

The release of the DOE report provided a timely platform for launching actions that DOE will take in response to energy system vulnerabilities. Next steps include work to better characterize response measures that might be taken to enhance climate preparedness and resilience of the energy system, and assess barriers in resilience investment decisions in the energy sector. DOE's efforts will help to coordinate other activities designed to enhance information, improve stakeholder coordination and inform on policies and other actions that could enhance deployment of climate-resilient energy technologies. This information will be helpful to DOE sites in conducting their vulnerability assessments and developing plans to address their vulnerabilities.

Grid Resilience

In 2013, DOE's Office of Electricity Delivery and Energy Reliability and the President's Council of Economic Advisers, with assistance from the White House Office of Science and Technology, released a report examining the economic impact of increasing the resilience of the U.S. electric grid[15]. The report estimates the annual cost of power outages caused by severe weather between 2003 and 2012 and describes various strategies for modernizing the grid and increasing grid resilience. Over this period, weather-related outages are estimated to have cost the U.S. economy an inflation-adjusted annual average of $18 billion to $33 billion. Grid resilience is increasingly important as climate change increases the frequency and intensity of severe weather.

Preparing for the challenges posed to the grid by climate change requires investment in 21st century technology. A multi-dimensional strategy will prepare the United States for climate change and the increasing incidence of severe weather. Developing a smarter, more resilient electric grid is one step that can be taken in the short term to develop energy stability that drives our economy into the future. DOE recognizes that it has an important role to play in this effort.

Increasing the Scientific Understanding of Climate Change

DOE's core mission to advance scientific understanding, as established in the *2014-2018 DOE Strategic Plan*, confirms the Department's commitment to the development of useful and timely decision-support tools for climate change preparedness. These efforts serve to meet the Department's contribution to the President's Climate Action Plan as well the goals outlined in the first installment of this plan. The Climate and Environmental Sciences Division (CESD), in the Office of Science's Office of Biological and Environmental Research, focuses on advancing a robust, predictive understanding of the Earth's climate and environmental systems that informs the development of sustainable solutions to the nation's energy and environmental challenges[16].

The division has three research activities, each of which directly informs climate science. The Climate and Earth System Modeling group's research seeks to develop high fidelity community models representing the coupled physical-human system, with a significant focus on understanding dynamical variability and system response to natural and anthropogenic forcing. Atmospheric System Research strives to understand the physics, chemistry, and dynamics governing clouds, aerosols, and precipitation interactions, with a goal to advance the predictive understanding of the climate system. The third research activity, Environmental System Science, focuses on advancing ecological and subsurface terrestrial science, with a goal to understand the full dynamics of the carbon cycle and how it influences climate change.

Numerous national laboratories within the Department are also involved in basic climate change science, climate modeling, and support tools. A number of these research facilities contributed to the International Panel on Climate Change (IPCC) Fifth Assessment Report and the Third National Climate Assessment. Examples of current research include:

- Pacific Northwest National Laboratory (PNNL) plays a leadership role in the multi-laboratory Atmospheric Radiation Measurement (ARM) Research Facility program, that in turn provides critical data necessary to reduce climate prediction uncertainty associated with the earth's radiation balance, with particular focus on clouds, aerosols, and precipitation systems.
- At the Lawrence Berkeley National Laboratory (LBNL), research is dedicated to understand the physical and stochastic nature of extreme events within the climate system, with a special focus to project how anthropogenic forcing of the climate leads to a redistribution of extremes, in both space and time.
- At the Oak Ridge National Laboratory (ORNL), significant research on decadal scale ecological field experiments, at both midlatitude and arctic sites. The ORNL activities in the Arctic, involve in particular a

network of numerous national laboratory and academic institutions, to understand the dynamics governing permafrost ecology and the future risks of abrupt climate change.

Future Activities

DOE is committed to its vision of mainstreaming and integrating climate change resiliency considerations across all DOE programs, wherever appropriate. To accomplish this goal, DOE will engage in a variety of activities in addition to those already ongoing.

Continuity of Operations Planning and Policy

DOE plans to update its Departmental Continuity of Operations (COOP) Plan in 2014 and will also look to consider climate change resilience. The COOP Plan is used to outline the actions taken to provide the capability to continue mission-essential processing and restore normal operations after a disaster or disruption. The inclusion of climatic risks into the COOP will enable DOE to increase resilience across multiple identified vulnerabilities. The update will be a joint-effort and supported by numerous offices throughout the Department, including the Office of Emergency Management located in the National Nuclear Security Administration (NNSA).

Infrastructure Resilience Working Group -Council on Climate Preparedness and Resilience

President Obama established the Council on Climate Preparedness and Resilience with Executive Order 13653, *Preparing the United States for the Impacts of Climate Change*. The Council will identify priority federal government actions related to climate change preparedness and resilience and coordinate agency and interagency efforts to implement those actions, including those described in the President's Climate Action Plan. Four working groups have been established to support the Council, including the Infrastructure Resilience Working Group, co- chaired by DOE and the Department of Homeland Security (DHS), with the focus of realizing innovative infrastructure resilience solutions to climate change. In addition to a high level assessment across critical infrastructure, the working group will pursue a "deep dive" on energy that includes: characterizing and prioritizing energy sector vulnerabilities and interdependencies to other sectors; identifying existing resilience activities, barriers, and key research and policy opportunities for enhancing resilience; and identifying metrics for measuring success. It will forge new interagency partnerships where appropriate, and undertake immediate action on multi-sector climate change preparedness and resilience efforts.

While this effort will look broadly across infrastructure, it will also target specific sectors such as the energy infrastructure. DOE and other agencies will leverage activities underway for the Quadrennial Energy Review (QER) and focus on vulnerabilities and climate change preparedness and resilience strategies for the energy sector. This work acknowledges the critical role energy will play, but will also identify interdependencies between energy and other sectors (e.g., communication, water, transportation, health, etc.). Specific tasks targeting energy infrastructure will include identifying federal energy technology R&D activities and policies/programs focused on climate resilience such as the update to the Quadrennial Technology Review, and identifying key gaps. In addition, the effort will identify relevant information, data, tools, best practices, and lessons learned for climate change preparedness and resilience for energy systems, as well as desired outcomes and measures for measuring progress towards energy sector preparedness and resilience to climate change. This information will be helpful as well to DOE sites in conducting their vulnerability assessments and developing plans to address their vulnerabilities.

Agency Procurement, Acquisition, and Real Property Decisions

DOE will consider the need to improve the integration of climate adaptation and resilience into procurement and real property decisions. The ability to address DOE physical asset and programmatic vulnerabilities facilitates the need for the Department to begin to contemplate such issues. DOE has already made climate adaptation planning a tenet of the Site Sustainability Plans and can leverage that mechanism to begin the discussion. DOE's pilot vulnerability assessments will look to transfer approaches and templates to other DOE sites. Further action requires widespread involvement of DOE procurement and facilities management offices.

DOE identified the Asset Management Plan as a possible conduit for incorporating resiliency into personal and real property planning. The Asset Management Plan is scheduled to be updated in FY 2014. The plan, combined with proper action, identification of climatic risk could allow DOE sites to more effectively manage assets in the short and long term. Another potential opportunity is DOE Order 430.1B, *Real Property Asset Management*, which includes requirements for preparing Ten Year Site Plans. The process of updating this order has not yet begun.

The DOE Acquisition Guide serves to supplement the Federal Acquisition Regulation (FAR) and the Department of Energy Acquisition Regulation (DEAR) by identifying relevant internal standard operating procedures to be followed by both procurement and program personnel who are involved in various aspects of the acquisition process. The guide is also intended to be a repository of best practices found throughout the Department. DOE will consider the inclusion of climate change resiliency to the guide as a way to ensure that challenges to procurement of critical inputs are addressed. DOE recognizes the threat climate change poses to acquisition activities due to supply chain disruption and will work to address this vulnerability.

Modernizing Federal Programs to Support Resilience Investments

E.O. 13653 calls for efforts to modernize federal programs and policies to support collaboration at local and regional levels. Departmental efforts to develop collaboration are largely addressed in actions to engage local and regional stakeholders on resiliency. The "Regional Collaboration and Interagency Cooperation" section of this Plan details current activities in support of this goal. DOE is in the process of evaluating appropriate policies, programs and plans. During these evaluations and potential updates DOE will look to identify barriers discouraging actions, reform programs inhibiting resilience, and identify opportunities to support smarter actions and investments. A particular focus will be paid to DOE's grants, loans and other financing programs but the effort will encompass DOE's entire mission.

DOE recognizes that its unique scientific and technologic innovation mission provides an opportunity to foster local and regional collaboration as well as encourage investment in resilience.

CONCLUSION

The 2014 Adaptation Plan is the second installment of an ongoing effort to build resilience across the Department. The plan serves as a foundation from which future updates will build. Future plans must account for advancement in scientific understanding and continued evaluation, made in accordance with E.O. 13653.

The 2014 Adaptation Plan will be incorporated into DOE's internal planning processes described above and through its integration with the goals of the annual SSPP, implemented by DOE's Senior Sustainability Officer, Sustainability Performance Office, and Senior Sustainability Steering Committee and their component DOE programs and sites.

Further, DOE will engage in and share best practices within the Department and with other federal agencies as appropriate through the Council on Climate Resilience, interagency working groups, and also by forging new collaborations with other agencies and stakeholders as appropriate. DOE will continue to leverage its unique modeling, climate science expertise, and engineering capabilities in collaboration with other agencies and institutions, to continuously improve understanding of the effects of climate change and identify appropriate adaptation strategies.

DOE will continue to include climate change adaptation as part of its planning and operations. A more resilient DOE enables the Department to maintain mission specific operations and minimize disruption to critical functions. A steadfast commitment to climate resilience strengthens Departmental operations by creating a more adaptive agency, and helps to create a resilient energy sector, and stronger, better prepared communities.

APPENDICES

Appendix 1

Table 1. Summary of Forecasted Climate Effects for Major US Climate Regions[17,18,19,20]

Region	Potential Effects of Greatest Concern to DOE
Alaska	• Longer summers and higher temperatures are causing drier conditions, even in the absence of strong trends in precipitation • Insect outbreaks and wildfires are increasing with warming • Lakes are declining in area • Thawing permafrost damages infrastructure (roads, runways, water and sewer systems, transmission lines, oil and gas pipelines, etc.) and increases the potential for the release of

Region	Potential Effects of Greatest Concern to DOE
	GHG gases (e.g., methane) from thawing permafrost • Coastal storms increase risks to coastal oil and gas facilities and operations
Great Plains	• Projected increases in temperature, evaporation, and drought frequency add to concerns about the region's declining water resources (impacts on steam electric power generation) • Agriculture (e.g., biofuels), and natural lands, already under pressure due to an increasingly limited water supply, are very likely to also be stressed by rising temperatures • Ongoing shifts in the region's population from rural areas to urban centers will interact with a changing climate, resulting in a variety of consequences to energy sector infrastructure and demand
Midwest	• During the summer, public health and quality of life, especially in cities, will be negatively affected by increasing heat waves and potential electricity shortages or brownouts, reduced air quality, and increasing insect and waterborne diseases. In the winter, warming will have mixed impacts • The likely increase in precipitation in winter and spring, more heavy downpours, and greater evaporation in summer would lead to more periods of both floods and water deficits affecting both hydropower, and steam electric power generation • While the longer growing season provides the potential for increased crop yields, increases in heat waves, floods, droughts, insects, and weeds will present increasing challenges to managing crops (e.g., for biofuels)
Northeast	• Extreme variability in heat and declining air quality are likely to pose increasing problems for human health, especially in urban areas • Agricultural production is likely to be adversely affected as favorable climates shift • Severe flooding due to sea-level rise and heavy downpours associated with more intense precipitation events is likely to occur more frequently • Significant sea-level rise and storm surge will adversely affect coastal cities, ecosystems, and energy infrastructure; low-lying and subsiding areas are most vulnerable

Table 1. (Continued)

Region	Potential Effects of Greatest Concern to DOE
Northwest	• Declining springtime snowpack leads to reduced summer stream flows, straining water supplies, and potentially reducing hydro-electric power generation • Increased insect outbreaks, wildfires, and changing species composition in forests will pose challenges for ecosystems and the forest products industry Ecosystems will experience additional stresses as a result of rising water temperatures and declining summer stream flows
Pacific Islands	• The availability of freshwater is likely to be reduced, with significant implications for island communities, economies, and resources • Island communities, infrastructure, and ecosystems are vulnerable to coastal inundation due to sea-level rise and coastal storms
Southeast	• Projected increases in air and water temperatures will cause heat-related stresses for people, plants, and animals • Decreased water availability is very likely to affect the region's economy, natural systems, and steam electric power generation • Sea-level rise and the likely increase in hurricane intensity and associated storm surge will impact energy infrastructure (e.g., oil and gas production and storage facilities) • Ecological thresholds are likely to be crossed throughout the region, causing major disruptions to ecosystems and to the benefits they provide to people • Quality of life will be affected by increasing heat stress, water scarcity, severe weather events, and reduced availability of insurance for at-risk properties
Southwest	• Water supplies will become increasingly scarce, calling for trade-offs among competing uses, (e.g., energy, agriculture, industry, domestic use, etc.) and potentially leading to conflict Increasing temperature, drought, wildfire, and invasive species will accelerate transformation of the landscape • Increased frequency and altered timing of flooding will increase risks to people, ecosystems, and infrastructure • Cities and human health systems face increasing risks from shifting disease vectors, temperature increases and health care system infrastructure from a changing climate

End Notes

[1] State of the Climate in 2012, NOAA http://journals.ametsoc.org/doi/suppl/10.1175/2013BAMSStateoftheClimate.1/supplfile/2013bamsstateoftheclimate.2.pdf

[2] 2013 DOE Strategic Sustainability Performance Plan http://www1.eere.energy.gov/sustainability/pdfs/doe sspp 2013.pdf

[3] 2014 DOE Strategic Plan

[4] USGCRP. Global Climate Change Impacts in the US. 2009: http://globalchange.gov/publications/reports/scientificassessments/us-impacts

[5] DOE High Level Analysis of Vulnerability to Climate Change, April 2012

[6] U.S. Department of Energy, Climate and Energy-Water-Land System Interactions: Technical Report to the U.S. Department of Energy in Support of the NCA, March 2012 http://www.pnnl.gov/main/publications/external/technical reports/PNNL21185.pdf

[7] U.S. Department of Energy, Effects of Climate Change on Federal Hydropower, August 2013 http://www1.eere.energy.gov/water/pdfs/hydro climate change report.pdf

[8] USCGCRP, 2013 NCA Regional Climate Scenarios Summaries Parts 1-9 http://scenarios.globalchange.gov/node/1155

[9] Burkett, Virginia and Margaret Davidson, "Coastal Impacts, Adaptation and Vulnerabilities: A Technical Input to the 2013 National Climate Assessment" NOAA, 2013 http://cakex.org/sites/default/files/documents/Coastal-NCA-1.13- web.form 0.pdf

[10] Federal Memorandum of Understanding Executive Order 12898 Federal Actions to Address Environmental Justice in Minority Populations and Low-Income Populations http://energy.gov/lm/services/environmental-justice

[11] National Climate Assessment – Health Sector Workshop: Southeast Region Report, 2012

[12] US Dept. of Energy, 2013 DOE Strategic Sustainability Performance Plan http://www1.eere.energy.gov/sustainability/pdfs/doe sspp 2013.pdf

[13] The State of the Future for a Sustainable Tennessee: Grand Challenges and Grand Opportunities Under a Changing Climate (http://sustainabletennessee.org/wp-content/uploads/2012/09/SustainableTN.pdf)

[14] Dept. of Energy Press Release June 13, 2013 http://energy.gov/articles/energy-department-partners-state-city-andindustry-stakeholders-help-hoboken-region-improve

[15] Economic Benefits of Increasing Electric Grid Resilience to Weather Outages, Executive Office of the President, August 2013 http://energy.gov/sites/prod/files/2013/08/f2/Grid%20Resiliency%20Report FINAL.pdf

[16] Climate and Environmental Sciences Division 2012 Strategic Plan http://science.energy.gov/~/media/ber/pdf/CESDStratPlan-2012.pdf

[17] USGCRP. Global Climate Change Impacts in the US. 2009. http://globalchange.gov/publications/reports/scientificassessments/us-impacts

[18] USCGCRP 2013 NCA Regional Climate Scenarios Summaries Parts 1-9 http://scenarios.globalchange.gov/node/1155

[19] U.S. Climate Change Science Program. Climate Change on Energy Production and Use in the United States. (Synthesis and Assessment Product 4.5). 2007. http://globalchange.gov/publications/reports/scientific-assessments/saps/sap4-5

[20] A. Anthony Bloom, Paul I. Palmer, Annemarie Fraser, David S. Reay and Christian Frankenberg. "Large-Scale Controls of Methanogenesis Inferred from Methane and Gravity Space Borne Data". Science Vol. 327. No. 5963. January 15, 2010: pp. 322-325. http://www.sciencemag.org/content/327/5963/322.full.pdf

In: U.S. Energy Infrastructure
Editor: Joanne R. Ballard
ISBN: 978-1-63482-286-2
© 2015 Nova Science Publishers, Inc.

Chapter 3

CLIMATE CHANGE SCIENCE: KEY POINTS*

Jane A. Leggett

SUMMARY

Though climate change science often is portrayed as controversial, broad scientific agreement exists on many points:

- The Earth's climate is warming and changing.
- Human-related emissions of greenhouse gases (GHG) and other pollutants have contributed to warming observed since the 1970s and, if continued, would tend to drive further warming, sea level rise, and other climate shifts.
- Volcanoes, the Earth's relationship to the Sun, solar cycles, and land cover change may be more influential on climate shifts than rising GHG concentrations on other time and geographic scales. Human-induced changes are super-imposed on and interact with natural climate variability.
- The largest uncertainties in climate projections surround feedbacks in the Earth system that augment or dampen the initial influence, or affect the pattern of changes. Feedback mechanisms are apparent in clouds, vegetation, oceans, and potential emissions from soils.

---

* This is an edited, reformatted and augmented version of a Congressional Research Service publication, CRS Report for Congress R43229, prepared for Members and Committees of Congress, from www.crs.gov, dated September 10, 2013.

- There is a wide range of projections of future, human-induced climate change, all pointing toward warming and associated sea level rise, with wider uncertainties regarding the nature of precipitation, storms, and other important aspects of climate.
- Human societies and ecosystems are sensitive to climate. Some past climate changes benefited civilizations; others contributed to the demise of some societies. Small future changes may bring benefits for some and adverse effects to others. Large climate changes would be increasingly adverse for a widening scope of populations and ecosystems.

As is common and constructive in science, scientists debate finer points. For example, a large majority but not all scientists find compelling evidence that rising GHG have contributed the most influence on global warming since the 1970s, with solar radiation a smaller influence on that time scale. Most climate modelers project important impacts of unabated GHG emissions, with low likelihoods of catastrophic impacts over this century. Human influences on climate change would continue for centuries after atmospheric concentrations of GHG are stabilized, as the accumulated gases continue to exert effects and as the Earth's systems seek to equilibrate.

The U.S. government and others have invested billions of dollars in research to improve understanding of the Earth's climate system, resulting in major improvements in understanding while major uncertainties remain. However, it is fundamental to the scientific method that science does not provide absolute proofs; all scientific theories are to some degree provisional and may be rejected or modified based on new evidence. Private and public decisions to act or not to act, to reduce the human contribution to climate change or to prepare for future changes, will be made in the context of accumulating evidence (or lack of evidence), accumulating GHG concentrations, ongoing debate about risks, and other considerations (e.g., economics and distributional effects).

BROAD SCIENTIFIC AGREEMENT ON MANY ASPECTS OF CLIMATE CHANGE[1]

Despite portrayals in popular media about controversies in climate change science, almost all climate scientists agree on certain important points:

- The Earth's climate has warmed significantly and changed in other ways over the past century (*Figure 1*). The warming has been widespread but not uniform globally, with most warming over continents at high latitudes, and slight cooling in a few regions, including the southeastern United States, the Amazon, and the North Atlantic south of Greenland.[2]

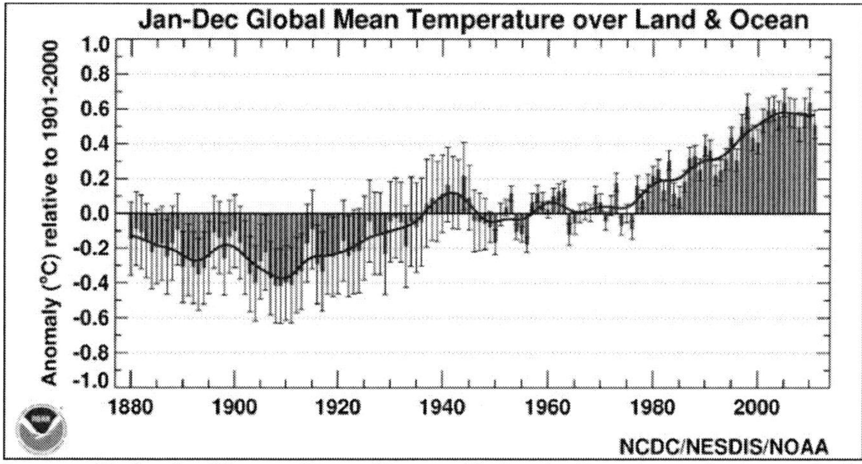

Source: National Climate Data Center, National Oceanic and Atmospheric Administration, U.S. Department of Commerce. Figure extracted March 28, 2013. Very similar findings are reported by several other, independent research groups. See, for example, Rohde, Robert, Richard Muller, Robert Jacobsen, Elizabeth Muller, Saul Perlmutter, Arthur Rosenfeld, Jonathan Wurtele, Donald Groom, and Charlotte Wickham. "A New Estimate of the Average Earth Surface Land Temperature Spanning 1753 to 2011." Geoinformatics & Geostatistics: An Overview 1, no. 1 (December 7, 2012).

Notes: Red bars represent "anomalies," or differences in mean temperature for the year compared with the 20th century average. Anomalies are a better estimate than the absolute value, as they can capture the change over time more reliably while absolute values are vulnerable to gaps in geographic coverage. The blue line shows the running average, applying a "21-point binomial filter" to the time series plotted as red bars. The "whisker" (thin black vertical) lines represent confidence or possible error levels. Confidence has improved over the past century.

Figure 1. Long-Term Temperature Observations. Compared to the 20th Century Global Mean Temperature.

- The climate has varied naturally through geologic history. Past climate changes sometimes proceeded abruptly when they passed certain "tipping points." The National Academy of Sciences concluded that the past few decades were very likely the warmest in the past 400 years, and "that temperatures at many, but not all, individual locations were higher during the past 25 years than during any period of comparable length since A.D. 900."[3] Although conclusions cannot yet be precise, research suggests that global average temperatures today are among the highest since human civilizations began to flourish roughly 4,000 to 8,000 years ago.[4]

- "Greenhouse gases" (GHG) include, among others, carbon dioxide (CO2), water vapor, methane (CH4), and nitrous oxide (N2O), as well as some aerosols. They absorb energy into the atmosphere rather than letting it escape to space. The presence of GHG in the atmosphere warms the Earth to its current temperature.
- Human activities, especially fossil fuel burning, deforestation, agriculture, and some types of industry, have increased GHG concentrations in the atmosphere. CO2, the main GHG emitted by human activities, has risen almost 40% over the past 150 years. About one-third of human-related CO2 has been absorbed by oceans, increasing surface water acidity by 30%.[5]
- The enhanced levels of GHG in the atmosphere contributed to the observed global average warming in recent decades. Over other time and geographic scales, such factors as the Earth's orbit, solar variability, volcanoes, and land cover change have been stronger influences than human-related GHG.
- There is a wide range of projections of future human-induced climate change, with all pointing toward warming. Human-induced change will be superimposed on, and interact with, natural climate variability.
- Human societies and ecosystems are sensitive to climate. Some climate changes benefited civilizations; others contributed to some societies' demises.
- The range of possible impacts on humans and ecosystems is also very wide. In the near term, climate change (including the fertilization of vegetation by CO2) may bring benefits for some, while adversely affecting others. Researchers expect the balance of projected climate change impacts to be increasingly adverse for a widening scope of populations and ecosystems.

As is common and constructive in science, scientists debate finer points. Some disagree with the broader consensus that GHG have been *the major* influence on global warming over the past few decades. Some suggest that, if GHG emissions continue unabated, the resulting climate change would be small and possibly beneficial overall. Most climate modelers project changes that are significant to large, with small likelihoods of changes that could be catastrophic for some human societies and ecosystems in coming decades.

DEALING WITH UNCERTAINTIES

Even the best science cannot provide absolute proof; it is fundamental to the scientific method that all theories are to some degree provisional and may be rejected or modified based on new evidence. Private and public decisions to act or not to act, to reduce the human contribution to climate change or to prepare for future changes, will be made in the context of accumulating evidence (or lack of evidence), accumulating GHG concentrations, ongoing debate about risks, and other considerations (e.g., economics and distributional effects). That said, billions of dollars have been invested in research on a wide range of climate change topics, including the possibility of attribution to alternative causes than greenhouse gases. To date, scientists have found little support for the hypothesis that GHG are not responsible for observed warming, nor have they found much evidence that other factors (including solar changes) can explain more than a small portion of global average temperature increases since the 1970s. For example, measurements of solar irradiance suggest that the solar influence on global temperatures has been decreasing overall since the 1930s, with the up-and-down pattern of the 11-year solar cycle evident in observations. A large body of research is consistent with attributing the majority of global temperature increase since the 1970s to the increase in GHG concentrations. It is this balance of evidence that leads most scientists to consider human-related GHG emissions an important global risk.

Sound Science Does Not Offer Proof

As scientists may point out, "there is no such thing as a scientific proof. Proofs exist only in mathematics and logic, not in science.... The primary criterion and standard of evaluation of scientific theory is evidence, not proof.... The currently accepted theory of a phenomenon is simply the best explanation for it among all available alternatives."6 Normal scientific methods aim at disproving a hypothesis; if evidence cannot disprove a hypothesis, it generally buttresses confidence in the hypothesis.

ISSUES FOR CONGRESS

It appears unlikely that science will provide decision-makers with significantly more scientific certainty for many years regarding the precise patterns and risks of climate change. Nonetheless, both private and public decision-makers face climate-related choices.

Broadly, response options to significant climate change include (1) defer the choices; (2) find out more; (3) inform affected populations; (4) prepare; (5) try to contain it; and (6) choose to experience the consequences. In many cases, many decision-makers are likely to face situations that require a response, such as resolving discrepancies between designated and actual flood plains or attempting to improve agricultural productivity in light of contemporary climate patterns.

Based on what is and what is not well known concerning climate change, as well as other considerations, Members of Congress may address climate-related decisions that affect

- authorizations and appropriations for federal programs, including research and technology development;
- tax and financial incentives for private decision-makers;
- regulatory authorities; or
- information or assistance to affected entities to help them adapt or rebuild after damages.

A variety of other CRS reports provide background and analysis on such options and are listed at the end of this report.[7]

Causes of Observed Climate Change: Forcings, Feedbacks, and Internal Variability

Three concepts may be useful for understanding the mechanisms and debate over the contributions to observed climate change: *forcings, feedbacks*, and *internal variability*.

Forcings

There is broad agreement among scientists that certain factors—including the composition of the atmosphere and solar variability—directly change the balance between incoming and outgoing radiation in the Earth's system and consequently *force* climate change. *Forcings* include the following:

- Atmospheric concentrations of greenhouse gas (GHG) and other trace gas and aerosol. These include water vapor,8 carbon dioxide (CO_2), methane (CH_4), nitrous oxide (N_2O), sulfur hexaflouride (SF6), chlorofluorocarbons (CFC), hydrofluorocarbons (HFC), perfluorocarbons (PFC), ozone, sulfur aerosols, black and organic carbon aerosols, and others. Human activities, especially fossil fuel burning, deforestation, agriculture, and some types of industry, have increased GHG concentrations in the atmosphere. CO_2 has risen almost 40% over the past 150 years.[9]
- Amount and patterns of solar radiation reaching the Earth due to the Earth's orbit around the Sun, and the tilt and wobble of the Earth's axis, as well as solar variability (Figure 2).
- Reflectivity of the Earth's surface due to changes in land use (e.g., urban surfaces, forest cover), changes in ice and snow cover; and vegetation cover.

Human-Related Greenhouse Gas Emissions

A majority of human-related GHG emissions are carbon dioxide, released primarily from energy production and use, deforestation and forest degradation, and cement manufacture. World-wide in 2010, carbon dioxide emissions were 74% of human-related GHG emissions. In the United States, carbon dioxide was 83% of human-related GHG emissions (*Figure 3*). Methane and nitrous oxide emissions are greater shares (16% and 8%, respectively) globally than in the United States (10% and 5%, respectively). Agriculture is a main source of these emissions and is a bigger share of the economies of many low-income countries compared with the United States. Also, many sources in the United States have acted to reduce their GHG emissions (such as in reducing leaks of methane), compared with sources in some low-income countries.

A large majority (73%) of global GHG are emitted by the 10 top emitting countries *Figure 4*). China became the leading GHG emitter in 2007 when it surpassed the United States. While China's emissions have been on the rise, the United States has emitted more cumulatively than any other country over the past 100 years.

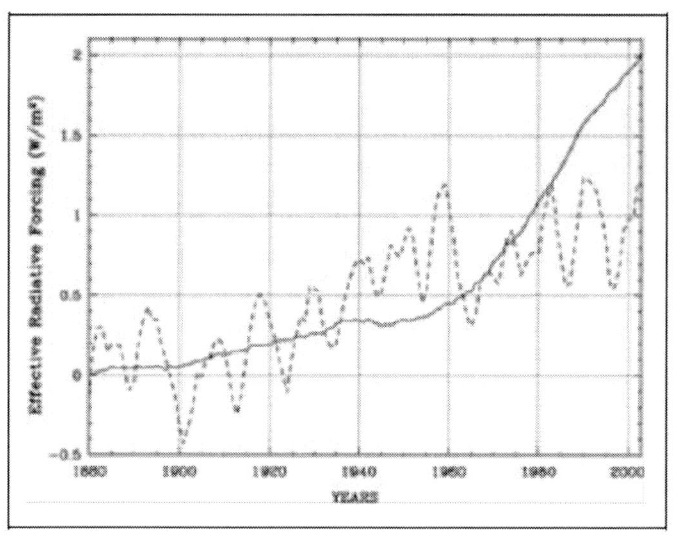

Source: Ziskin, Shlomi, and Nir J. Shaviv. "Quantifying the Role of Solar Radiative Forcing over the 20th Century." Advances in Space Research 50, no. 6 (September 15, 2012): 762–776. doi:10.1016/j.asr.2011.10.009.

Notes: According to the authors, "the optimal anthropogenic contribution (solid line) and the optimal solar contribution (dashed line) over the 20th century. The anthropogenic contribution is primarily composed of GHGs and aerosols. The solar contribution includes changes in the total solar irradiance and the indirect solar effect (ISE)." This is one of many studies, using a variety of methods, investigating the relative contributions of different climate forcings that conclude that the GHG concentrations have outweighed all other influences on global mean air surface temperature from the late 1970s to the present. For a broader, more thorough review of scientific understanding of the solar influence, see Gray, L. J., J. Beer, M. Geller, J. D. Haigh, M. Lockwood, K. Matthes, U. Cubasch, et al. "Solar Influences on Climate." Reviews of Geophysics 48, no. 4 (October 30, 2010). doi:10.1029/2009RG000282.

Figure 2. One Estimate of Human-Related versus Solar Contributions to Global Temperature Change Over the 20th Century.

Feedbacks

Once a change in the Earth's climate system is underway, responses *within* the system will amplify or dampen the initiated change. Virtually all climate scientists conclude that all the feedbacks *in net* are likely to be positive (i.e., increasing climate change in the same direction caused by warming),[10] especially if temperature increases are large;[11] there remain wide differences in views.

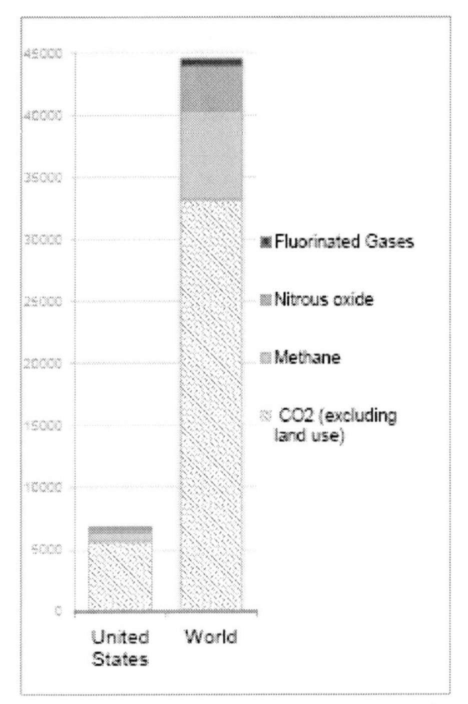

Source: CRS figure using estimates from World Resources Institute, CAIT version 2.0, extracted September 12, 2013.

Notes: These estimates cover the six GHG covered by the Kyoto Protocol (CO_2, CH_4, N_2O, SF6, HFC, and PFC), expressed in their equivalencies to the effect of CO_2 on "radiative forcing" of the atmosphere over a 100-year period.

The World column includes U.S. emissions.

Figure 3. Shares of Human-Related GHG Emissions by Gas in 2010. Million metric tons of CO_2-equivalent.

An important consideration is that, once positive feedbacks begin, they may be essentially irreversible and, at least theoretically, lead to "runaway warming." A few of the major feedbacks are clouds, vegetation, snow and ice cover, and uptake or releases of GHG by soils and water bodies. Forests, for example, provide both negative and positive feedbacks. On the one hand, higher CO2 concentrations in the atmosphere tend to fertilize their growth (if other conditions are not limiting) and forests may grow more rapidly with greater warmth and precipitation; these factors could dampen initiated warming. On the other hand, forests thrive within certain bounds of growing conditions; if their climate conditions change beyond those bounds, they are likely to grow more slowly and eventually die back, releasing the carbon they and forest soils store and enhancing the initiated climate change.

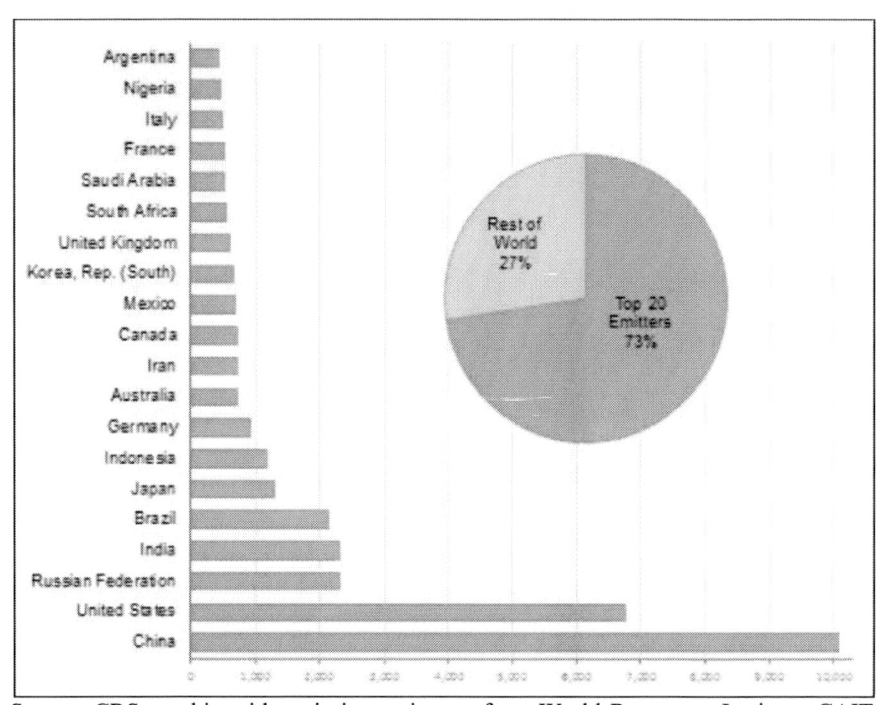

Source: CRS graphic with emission estimates from World Resources Institute, CAIT Version 2.0, extracted September 12, 2013.

Notes: These estimates cover the six GHG covered by the Kyoto Protocol (CO_2, CH_4, N_2O, SF6, HFC, and PFC), expressed in their equivalencies to the effect of CO_2 on "radiative forcing" of the atmosphere over a 100- year period.

Figure 4. Estimated Top 20 Emitting Nations of Greenhouse Gases in 2010. Million metric tons of carbon-dioxide equivalent, includes all net land use fluxes.

Internal Variability

The climate exhibits its own rhythms, or *internal variability*. The oscillation between El Niño and La Niña events is an example of internal climate variability that has important effects on economies and ecosystems in the Pacific basin (including across the United States). Another is the North Atlantic Oscillation. Internal variability may be difficult to distinguish from decadalscale climate change. Such patterns of variability also may be influenced by climate change.

PROJECTIONS OF FUTURE HUMAN-INDUCED CLIMATE CHANGE

Most climate science experts project that if GHG emissions are not reduced far below current levels, the Earth's climate would warm further, above natural variations, to levels never experienced by human civilizations. If, and as, the climate moves further from its present state, it would reconfigure the patterns and events to which current human and ecological systems are adapted, and the risk of abrupt changes would dramatically increase.

Scenarios of future GHG concentrations under current policies range from 500 ppm carbon dioxide equivalents[12] (CO_2e) to over 1,000 ppm CO_2e by 2100. These are projected to raise the global average temperature by at least 1.5° Celsius (2.7° Fahrenheit) above 1990 levels,[13] not taking into account natural variability. The estimates considered most likely by many scientists are for GHG-induced temperature increases around 2.5 to 3.2° C (4.5 to 5.8° F) by 2100.[14] There is a small but not trivial likelihood that the GHG-induced temperature rise may exceed 6.4°C (11.5° F) above natural variability by 2100.[15]

As context, the global average temperature at the Last Glacial Maximum has been estimated to be about 3 to 5° C (5.4 to 9° F) cooler than present,[16] and is estimated currently to be approaching the highest level experienced since the emergence of human civilizations about 8,000 years ago.[17]

Future climate change may advance relatively smoothly or sporadically, and some regions are likely to experience more fluctuations in temperature, precipitation, and frequency or intensity of extreme events than others. Almost all regions are expected to experience warming; some are projected to become warmer and wetter, while others would become warmer and drier. Sea levels could rise due to ocean warming alone on average between 7 and 23 inches by 2100. Adding to that estimate would be the effects of poorly understood but possible accelerated melting of the Greenland or Antarctic ice sheets. Recent scientific studies have projected a total global average sea level to rise in the 21st century, depending on GHG scenarios, ice dynamics, and other factors, in the range of 2 to 2.5 feet, with a few estimates ranging up to 6.5 feet.[18] Continued warming could lead to additional sea level rise over subsequent centuries of several to many meters. Improving understanding of ice dynamics is a high priority for scientific research to improve sea level rise projections.

Patterns consistent among different climate change models have led to some common expectations: GHG-induced climate change would include more heat waves and fewer extreme cold episodes; more precipitation on average but more droughts in some regions; and generally increased summer warming and dryness in the central portions of continents. Regional changes may vary from the global average changes, however. Scientists also expect precipitation to become more intense when it occurs, thereby increasing runoff and flooding risks.

Precipitation is a particularly challenging component of projecting future climate. For example, for the contiguous United States, recent climate modeling consistently anticipates overall temperature increases, but different models produce a wide range of precipitation changes, from net decreases to net increases.[19] This is particularly problematic in that precipitation, and its characteristics, is closely associated with impacts on agriculture, water supply, streamflows, and other critical systems.

Scientific expectations and model projections consistently point to a global average increase in precipitation with strong variations across regions and time. Generally, dry areas are expected to get dryer, and wet regions are expected to get wetter. In many regions, the increase in evapotranspiration is expected to exceed the increase in precipitation, resulting in general drying of soils and increasing risks of droughts. Precipitation, when it occurs, is expected to be more intense. There will be more energy available for storms, including hurricanes and thunderstorms, though whether they may increase in frequency remains unclear. Sea ice cover in the Arctic is projected to continue its recent decline (*Figure 5*). Greenland is expected to continue ice loss, adding to sea level rise, with more uncertainty about what may happen to ice cover in Antarctica (*Figure 5*). Because Arctic sea ice already floats on water, its melting would not increase sea levels, but large scale melting of land-based ice in Greenland and Antarctica could increase average sea levels by as many as 2 meters by 2100 and several more meters over coming centuries.

IMPACTS OF CLIMATE CHANGE

Nearly every human and natural system could be affected by climate changes, directly or indirectly. The U.S. Global Change Research Program has produced several assessments of scientific understanding of impacts of climate change on the United States.[20]

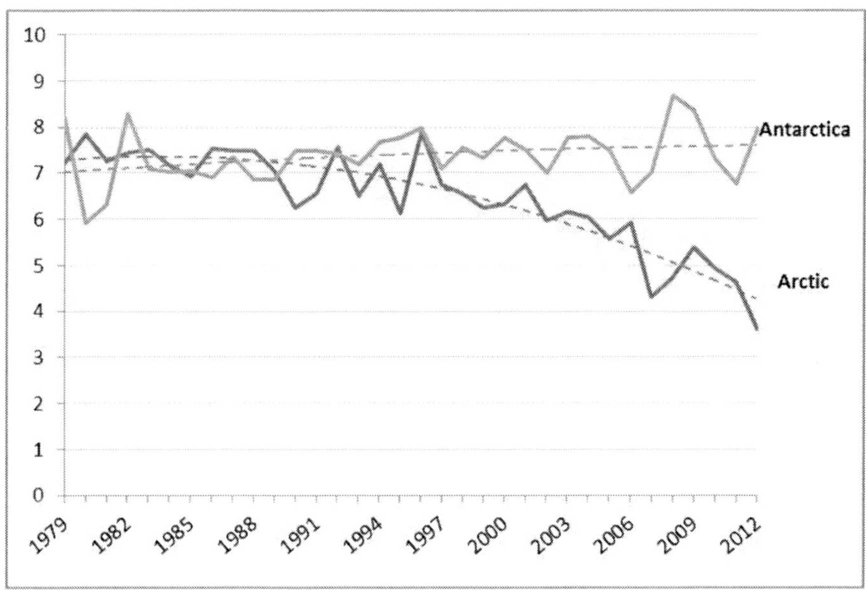

Source: CRS figure from data at the National Snow and Ice Data Center (extracted March 29, 2013), at http://nsidc.org/data/docs/noaa/g02135_seaice_index/ # monthly_graphs_format.

Notes: The dotted lines show the best polynomial fit (2-order) to each time series, as estimated by Excel. For both series, the polynomical fit was slightly better than a linear fit. See the National Snow and Ice Data Center website, referenced above, for further description of the underlying data.

Figure 5. Sea Ice Extent in Arctic (September) and Antarctica (April), 1979 to 2012. (in millions of square kilometers)

Climate Changes Would Affect A Wide Set of Human Systems

Changes in patterns of temperature, precipitation, sea levels, storms, and heat waves (among other indicators of climate) would affect, among other systems:

- water resources and delivery;
- agricultural productivity;
- the frequency and intensity of extreme weather events;
- spread of infectious diseases; air and water pollution levels;
- reliability of transportation, energy, and coastal protection systems;

- commodity prices;
- insurance pay-outs; and
- migration of people and species.

There are many additional elements of the economy and society that could be affected by shifts in climate. Research on potential impacts of climate change is generally less funded and developed than on the climate system itself.

Whether climate changes are meaningful in a policy context arguably depends, on the one hand, on how they influence existing and emerging human systems, and on the other hand, the values people attach to different resources and risks. Past climate changes, often regional not global, contributed to major societal changes, including some large-scale migrations and even the demise of some civilizations.[21] Some climate changes likely stimulated technological advances, such as development of irrigation systems.

Many investments in current buildings, transportation, water systems, agricultural hybrid varieties, and other infrastructure were designed in the context of a climate of one or more decades ago, cooler on average than today. To the degree that climate patterns were factored into design, the investments typically presumed that climate would remain stable within historical bounds of variability. For example, levees may have been built to withstand a 100-year flood (1% chance to occur each year) according to historic runoff, streamflow, and storm surge conditions going back many decades. As climate changes produce greater, more intense precipitation and run-off, however, a 100-year flood may now be closer to the 50-year flood (2% chance flood), and potentially the 10-year flood within decades (10% change flood). If climate continues to change from the conditions for which infrastructure and practices were designed, the risks of losses due to maladaptation would increase.

A wide band of uncertainty surrounds projections of impacts of climate change and, in particular, the critical thresholds for non-linear or abrupt effects. Some impacts of climate change are expected to be beneficial in some locations with a few degrees of warming (e.g., increased agricultural productivity in some regions, less need for space heating, less cold weather mortality, opening of the Northwest Passage for shipping and resource exploitation). Most impacts are expected to be adverse (e.g., lower agricultural productivity in many regions, drought, rising sea levels, spread of disease vectors, greater needs for cooling). Many impacts may be substantial but hard to assess as yet as "positive" or "negative," such as possible impacts on the structure of global financial markets. Risks of abrupt, surprising climate changes with accompanying dislocations are expected to increase as global average temperature increases; some could push natural and socioeconomic systems past key thresholds of tolerance.

Risks of future climate change would be reduced by efforts that reduce vulnerability and build resilience ("adaptation"). Some populations will have the resources to migrate and adapt successfully—even profit from new opportunities that will emerge—while others could lose livelihoods or lives. Adaptations can help mitigate impacts and damage costs, but also impose costs, often on those who can least afford them. Climate change will occur with different magnitudes and characteristics in different regions. The difficulties involved in improving predictions at regional and local scales will challenge preparations for climate change. To a large degree, climate change will expand the uncertainties that individuals and organizations face.

Climate change could have a wide array of effects on individuals, communities, and populations on a large scale. Many of these are expected to occur in small increments: shortages and increasing prices for clean water, rising food prices, higher rates of allergies and such illnesses as diarrhea or cholera, erosion of beaches, etc.

At an increasing rate may be shocks, or distinct weather events, such as more extreme heat waves, severe droughts, or loss of industrial cooling systems when intake water is in short supply or is warmer than tolerable temperatures.[22]

Atmospheric Carbon Dioxide is Increasing Ocean Acidity

The *acidity of the surface waters of the oceans* has increased by about 26% over the past 150 years.[23] Ocean acidification has occurred along with the rise in atmospheric concentrations of CO_2. The oceans remove 25%-40% of the carbon dioxide emissions added annually to the atmosphere by burning fossil fuels. The carbon dioxide absorbed in the oceans decreases the water's pH, an indicator of increasing acidity. According to a National Research Council (NRC) report, the current rate of acidification "exceeds any known change in ocean chemistry for at least 800,000 years."[24] Research shows varying sensitivities of different marine species to rising acidity, making general statements about impacts of ocean acidification difficult.

The NRC concluded,

> While the ultimate consequences are still unknown, there is a risk of ecosystem changes that threaten coral reefs, fisheries, protected species, and other natural resources of value to society. (*Executive Summary*, pp. 3-4)
>
> Congress enacted the Federal Ocean Acidification Research and Monitoring Act of 2009 (P.L. 111-11, Section 12311, Subtitle D) to improve monitoring and research, to assess carbon storage in the oceans and potential effects on acidification and other ocean conditions, and to develop predictive models for future changes in ocean chemistry and marine ecosystems. The program is housed within the National Oceanic and Atmospheric Administration (NOAA), and coordinated with other agencies through an interagency plan through the National Ocean Council.

Extreme events, chronic economic losses, or improved opportunities elsewhere are expected to prompt migration of millions of people, largely within countries, but also across national borders. Extreme events, greater variability, and uncertainty are expected to increase stress and mental health challenges. Some experts project that climate changes could amplify instabilities in countries with weak governance and increase security risks.[25] This may have implications for international political stability and security.[26]

For some experts and stakeholders, likely ecological disruptions (and limitations on species' and habitats' abilities to adapt at the projected rate of climate change) are among the most compelling reasons that humans must act to reduce their interference with the climate system. Some believe humans will have the wherewithal to cope, but non-human systems may not. As the degree and distribution of climate changes continue, ranges of species are likely to change. Climate change is highly likely to create substantial changes in ecological systems and services[27] in some locations, and may lead to ecological surprises.[28] The disappearance of some types of regional ecosystems raises risks of extinctions of species, especially those with narrow geographic or climatic distributions, and where existing ecological communities disintegrate.[29] One set of researchers found "a close correspondence between regions with globally disappearing climates and previously identified biodiversity hotspots; for these regions, standard conservation solutions (e.g., assisted migration and networked reserves) may be insufficient to preserve biodiversity."[30]

SELECTED, RELATED CRS REPORTS

CRS Report R43185, *Ocean Acidification*, by Harold F. Upton and Peter Folger.

CRS Report R41153, *Changes in the Arctic: Background and Issues for Congress*, coordinated by Ronald O'Rourke.

CRS Report RL34580, *Drought in the United States: Causes and Issues for Congress*, by Peter Folger, Betsy A. Cody, and Nicole T. Carter.

CRS Report R43199, *Energy-Water Nexus: The Energy Sector's Water Use*, by Nicole T. Carter.

CRS Report R42611, *Oil Sands and the Keystone XL Pipeline: Background and Selected Environmental Issues*, coordinated by Jonathan L. Ramseur.

CRS Report R42756, *Energy Policy: 113th Congress Issues*, by Carl E. Behrens.

CRS Report R42613, *Climate Change and Existing Law: A Survey of Legal Issues Past, Present, and Future*, by Robert Meltz.

CRS Report R43120, *President Obama's Climate Action Plan*, coordinated by Jane A. Leggett. CRS Report R42756, *Energy Policy: 113th Congress Issues*, by Carl E. Behrens. CRS Report RL34266, *Climate Change: Science Highlights*, by Jane A. Leggett.

CRS Report R41973, *Climate Change: Conceptual Approaches and Policy Tools*, by Jane A. Leggett.

End Notes

[1] This CRS report will be reviewed and, as appropriate, revised considering evidence provided from emerging scientific research. Of, note, the Intergovernmental Panel on Climate Change (IPCC) will release its fifth assessment report later in 2013.

[2] For more information, see maps available at the National Climate Data Center, http://www.ncdc.noaa.gov/oa/climate/globalwarming.html and http://www. ncdc.noaa.gov /oa/climate/research/trends.html#global.

[3] Board on Atmospheric Sciences and Climate. Surface Temperature Reconstructions for the Last 2,000 Years. National Research Council, 2006. http://books.nap.edu/openbook.php? record_id=11676&page=1.

[4] Marcott, Shaun A., Jeremy D. Shakun, Peter U. Clark, and Alan C. Mix. "A Reconstruction of Regional and Global Temperature for the Past 11,300 Years." Science 339, no. 6124 (March 8, 2013): 1198–1201. doi:10.1126/science.1228026; Kellerhals, T., S. Brütsch, M. Sigl, S. Knüsel, H. W. Gäggeler, and M. Schwikowski. "Ammonium Concentration in Ice Cores: A New Proxy for Regional Temperature Reconstruction?" Journal of Geophysical Research: Atmospheres 115, no. D16 (2010): n/a–n/a. doi:10.1029/2009JD012603; Thibodeau, Benoît, Anne de Vernal, Claude Hillaire-Marcel, and Alfonso Mucci.

"Twentieth Century Warming in Deep Waters of the Gulf of St. Lawrence: A Unique Feature of the Last Millennium." Geophysical Research Letters 37, no. 17 (2010): n/a–n/a. doi:10.1029/2010GL044771. See also the references at http://www.globalwarmingart.com/wiki/File:Holocene_Temperature_Variations_Rev_png, which depict a collection of major temperature reconstructions of the Holocene, as well as the broad range of uncertainty of available estimates and the average of those estimates.

[5] National Research Council. Ocean Acidification: A National Strategy to Meet the Challenges of a Changing Ocean. Washington DC, 2013; Feely, Richard A. 2010. A Rational Discussion of Climate Change: The Science, the Evidence, the Response. Testimony before the House Committee on Science and Technology, Subcommittee on Energy and Commerce. Washington DC. (p.130). See also CRS Report R43185, Ocean Acidification, by Harold F. Upton and Peter Folger.

[6] See, for example, the discussion in Kanazawa, Satoshi. "Common misconceptions about science I: "Scientific proof"." Psychology Today, November 16, 2008. http://www.psychologytoday.com/blog/the-scientific-fundamentalist/ 200811/common-misconceptions-about-science-i-scientific-proof.

[7] Many CRS reports related to climate change may be found at Issues Before Congress: Climate Change Science, Technology, and Policy, at http://www.crs.gov/pages/subissue.aspx?cliid=3878&parentid=2522&preview=False.

[8] Water vapor is the most important GHG in the atmosphere but is understood not to be directly influenced by humans; it would be, however, involved in feedback mechanisms, discussed later.

[9] About one-third of human-related CO2 has been absorbed by oceans, increasing surface water acidity by 30%. See National Research Council. Ocean Acidification: A National Strategy to Meet the Challenges of a Changing Ocean. Washington, DC, 2013; Feely, Richard A. 2010. A Rational Discussion of Climate Change: The Science, the Evidence, the Response. Testimony before the House Committee on Science and Technology, Subcommittee on Energy and Commerce. Washington, DC. (p.130).

[10] One line of evidence is that carbon dioxide levels have varied closely with the Earth's temperature in and out of glacial periods over the past million years. These cycles are mostly triggered by changes in the Earth's orbit, tilt, and wobble. In some of these cycles, temperatures rose in advance of rising atmospheric carbon dioxide concentrations. Scientists generally interpret this as a tendency for positive climate warming feedbacks that increase carbon dioxide concentrations which then enhance warming, etc.—that the net positive feedbacks amplify an initial climate warming.

[11] Positive feedbacks could increase if and when, for example, large tracts of forests die back as a response to exceeding their climate thresholds of tolerance, or current permafrost thaws and releases the carbon it contains, or if reservoirs of methane hydrates destabilize.

[12] In order to show multiple gases of different potencies on a single scale, GHG have been indexed relative to the effect that a mass of CO2 would have over several time periods (because GHG remain in the atmosphere for different lengths of time, from days to tens of thousands of years). The index used for these estimates uses a 100-year time horizon, the most frequently used period.

[13] Intergovernmental Panel on Climate Change Working Group I, Climate Change 2007: The Physical Basis (Cambridge, UK: Cambridge University Press, 2007).

[14] As a point of reference, the global mean annual temperature during the 20th century is estimated to have been approximately 13.9o Celsius (57.0o Fahrenheit), according to NOAA's National Climate Data Center.

[15] Ibid.

[16] Intergovernmental Panel on Climate Change Working Group I. Climate Change 2007: The Physical Basis. Cambridge, UK: Cambridge University Press, 2007. Executive Summary.

[17] Highest temperatures of the Holocene may have occurred in one or more periods some 5,000 to 8,000 years ago, although sufficient data are not available for all parts of the globe to have reliable estimates of average global temperature. The oldest cities discovered date from approximately the same period, such as the extensive settlement of Byblos in present-day Lebanon, by about 6,000 years ago, or Medinat Al-Fayoum in Egypt, about 6,000 years old. Since the early to mid-Holocene, however, average temperatures appear to have been declining slowly, with notable periods of warming and cooling. The changes entailed in Holocene climate variability have been significant in terms of effects on humans and ecosystems, and have led to both benefits to, and the demise of, numerous civilizations.

[18] See discussion in National Research Council. Advancing the Science of Climate Change. Washington DC, 2010, at p. 244.

[19] See, for example, climate change scenarios available from the U.S. Global Change Research Program at http://scenarios.globalchange.gov/sites/default/ files/b/figures/ UnitedStates/ Ann_US_precip_a2.png, with notes at http://scenarios.globalchange.gov/node/1087. See also discussion in this report regarding dealing with uncertainties.

[20] Karl, Thomas R., Mellillo, Jerry M., and Peterson, Thomas C. (eds.) Global Change Impacts in the United States. U.S. Global Change Research Program. 2009. Such periodic assessments are required by the Global Change Research Act of 1990 (P.L. 101-606). A new national assessment of impacts on the United States is due in late 2013.

[21] There is a growing set of research on the relationship between past climate change and civilizations. A sample of recent research includes Buckley, B. M., K. J. Anchukaitis, D. Penny, R. Fletcher, E. R. Cook, M. Sano, L. C. Nam, A. Wichienkeeo, T. T. Minh, and T. M. Hong. "Climate as a Contributing Factor in the Demise of Angkor, Cambodia." Proceedings of the National Academy of Sciences 107, no. 15 (March 2010): 6748–6752; Cook, Edward R, Kevin J Anchukaitis, Brendan M Buckley, Rosanne D D'Arrigo, Gordon C Jacoby, and William E Wright. "Asian Monsoon Failure and Megadrought During the Last Millennium." Science (New York, N.Y.) 328, no. 5977 (April 23, 2010): 486– 489; DeMenocal, PB. "Cultural Responses to Climate Change During the Late Holocene." Science (Washington) 292, no. 5517 (April 27, 2001): 667–673; Haug, G. H, D. Gunther, L. C Peterson, D. M Sigman, K. A Hughen, and B. Aeschlimann. "Climate and the Collapse of Maya Civilization." Science 299, no. 5613 (2003): 1731; Scholz, Christopher A., Thomas C. Johnson, Andrew S. Cohen, John W. King, John A. Peck, Jonathan T. Overpeck, Michael R. Talbot, et al. "East African Megadroughts Between 135 and 75 Thousand Years Ago and Bearing on Early-modern Human Origins." Proceedings of the National Academy of Sciences of the United States of America 104, no. 42 (October 16, 2007): 16416–16421.

[22] For examples of these risks to power plants, see Department of Energy (DOE), U.S. Energy Sector Vulnerabilities to Climate Change and Extreme Weather. July 2013. http://energy.gov/sites/prod/files/2013/07/f2/20130710-EnergySector-Vulnerabilities-Report.pdf. See also CRS Report R43199, Energy-Water Nexus: The Energy Sector's Water Use, by Nicole T. Carter.

[23] NRC Committee on the Development of an Integrated Science Strategy for Ocean Acidification Monitoring, Research, and Impacts Assessment; National Research Council. "Executive Summary." In Ocean Acidification: A National Strategy to Meet the Challenges of a Changing Ocean. Prepublication. Washington, D.C.: The National Academies Press, 2010; Jacobson, Mark Z. "Studying ocean acidification with conservative, stable numerical

schemes for nonequilibrium air-ocean exchange and ocean equilibrium chemistry." Journal of Geophysical Research 110 (April 2, 2005): 17 PP.

[24] Ibid.

[25] An example of this is the adverse weather events in early 2011 that led to spikes in food prices and contributed to demonstrations in Tunisia and Egypt. These, in turn, led to regime change, although one cannot attribute these events to climate change, as opposed to weather variability, and the political implications might have been very different in regimes with better economic performance, less income disparity, fewer allegations of corruption, and greater social resilience. The point remains, nonetheless, that societies are sensitive to climatic variables in many ways.

[26] Regarding risks to national security, see, for example, Defense Science Board Task Force on Trends and Implications of Climate Change for National and International Security. October 2011. http://www.acq.osd.mil/dsb/ reports/ADA552760.pdf; and U.S. Department of Defense. Quadrennial Defense Review, February 2010. (pp. xv, 84- 88) http://www.defense.gov/qdr/qdr%20as%20of%2029jan10%201600.pdf.

[27] Economists and scientists sometimes refer to "ecosystem services," which are the services that natural systems provide and for which, very frequently, humans do not typically pay. Ecosystems services include water filtration, filtering of air pollution, recreational and spiritual opportunities, etc. Even without being valued in capital markets, ecosystem services may be critically important to economies. For example, in many coastal areas, mangroves or wetlands provide valuable buffering against frequent storm and flood events. If such ecosystem services did not exist, communities would have to pay for manufactured alternatives (e.g., sea walls) or risk incurring damages.

[28] For example, the very rapid spread of pine beetles in recent years was unexpected and caused large damages (although a temporarily inexpensive supply of timber) in a very short period. See CRS Report R40203, Mountain Pine Beetles and Forest Destruction: Effects, Responses, and Relationship to Climate Change.

[29] Malcolm, Jay R., Canran Liu, Ronald P. Neilson, Lara Hansen, and Lee Hannah. "Global Warming and Extinctions of Endemic Species from Biodiversity Hotspots." Conservation Biology 20, no. 2 (2006): 538-548.

[30] John W. Williams, Stephen T Jackson, and John E. Kutzbach, "Projected distributions of novel and disappearing climates by 2100 AD," Proceedings of the National Academy of Sciences of the United States of America 104, no. 14 (April 3, 2007).

INDEX

#

A

B

E

F

G

H

I

J

K

L

M

N

O

P

Q

R

S

T

U

V

W